THE GREAT DINOSAUR DISCOVERIES

University of California Press, one of the most distinguished university presses in the United States, enriches lives around the world by advancing scholarship in the humanities, social sciences, and natural sciences. Its activities are supported by the UC Press Foundation and by philanthropic contributions from individuals and institutions. For more information, visit www.ucpress.edu.

University of California Press
Berkeley and Los Angeles, California

Library of Congress Cataloging-in-Publication Data
Naish, Darren.
    The Great Dinosaur Discoveries / Darren Naish.
        p.   cm.
    Includes bibliographical references and index.
    ISBN 978-0-520-25975-1 (cloth : alk. paper)
    1. Dinosaurs.   2. Paleontology—Mesozoic.   3. Discoveries in science.   I. Title.

QE861.4.N35   2009
567.9—dc22   2009006140

18 17 16 15 14 13 12 11 10 09
10 9 8 7 6 5 4 3 2 1

Conceived, edited, and designed
in the United Kingdom by
Marshall Editions
The Old Brewery
6 Blundell Street
London N7 9BH
www.marshalleditions.com

Managing Editor Paul Docherty
Art Director Ivo Marloh
Picture Manager Veneta Bullen
Layout 3RD-I, Vanessa Green, Cecilia Bandiera
Copy Editor Ben Hoare
Indexer Sue Butterworth
Production Nikki Ingram

Originated in Hong Kong by Modern Age.
Printed and bound in Singapore by Star Standard Industries (Pte) Ltd.

Previous page *Stegosaurus* was discovered in the rocks of the Morrison Formation in the United States, and the Morrison remains a rich source of dinosaurs even today. Here, Dr. Robert Bakker and colleagues excavate a Morrison site at Como Bluff, Wyoming, in search of dinosaurs and other fossils.

Facing page *Acrocanthosaurus* was a giant predator that lived across the United States during the late part of the Early Cretaceous, between 115 and 100 million years ago. It reached 40 ft (12 m) in length and may have hunted the far bigger herbivore *Sauroposeidon*, seen in the background.

A Note on Terminology
Units of geological time are qualified by the terms terms "Early," "Late," and, sometimes, "Middle." *Tyrannosaurus* lived, for example, during the Late Cretaceous. Rock layers that correspond to these periods of geological time are qualified by the terms terms "Upper," "Lower," and, sometimes, "Middle": *Tyrannosaurus* was discovered in the Upper Cretaceous Hell Creek Formation.

THE GREAT DINOSAUR DISCOVERIES

Darren Naish

UNIVERSITY OF CALIFORNIA PRESS

BERKELEY   LOS ANGELES

Contents

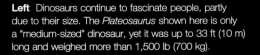

Left Dinosaurs continue to fascinate people, partly due to their size. The *Plateosaurus* shown here is only a "medium-sized" dinosaur, yet it was up to 33 ft (10 m) long and weighed more than 1,500 lb (700 kg).

Introducing Dinosaurs

When William Buckland and Gideon Mantell described the first fragmentary dinosaur bones during the 1820s, they could scarcely have imagined the scale of the discoveries that would follow in centuries to come. More so than any other group of animals in history, dinosaurs have captured the public's attention, and children and adults alike are fascinated by their sizes and remarkable shapes. Today, the remains of dinosaurs are often the most popular exhibits in museums, and hardly a week goes by when dinosaurs are not featured in the global news media.

Facing page The gigantic skull of the predatory dinosaur *Carcharodontosaurus* dwarfs a human skull. The size and imagined ferocity of some dinosaurs explains their appeal to the public. How remarkable animals such as this behaved when alive, how they changed over time, and how they are related to one another and to other animals, makes them the source of enduring fascination for scientists too.

What this book is about

Most dinosaur books look at current views on dinosaurs, and briefly recap the history of some key finds; some mention that old views on a given dinosaur were quite different from modern views. This book is specifically focused on changing ideas about the evolution and appearance of dinosaurs and the important discoveries that brought about these changes.

The history of dinosaur science breaks down into several stages. The first discoveries were made in the 1820s and 30s, and during the 1840s the concept of dinosaurs was conceived. During this early phase, experts considered the implications dinosaurs had for the Biblical story of creation and whether they provided evidence for evolution. They compared what few dinosaur bones they had to those of other creatures and struggled to imagine these strange beasts as live animals. During the late 1800s, exploration in North America resulted in a golden age of discovery. Most of the dinosaur skeletons and replicas mounted in museums originate from this time, and the discovery of good, articulated skeletons allowed scientists to reconstruct musculature and imagine ranges of movement. Several paleontologists continued working on dinosaurs into the 1920s and 30s, but efforts then slowed considerably as research focus moved elsewhere. Dinosaur research then entered a long quiet phase that lasted until the 1960s.

At the end of the 1960s, new discoveries and ideas resulted in a burst of interest known as the Dinosaur Renaissance. New techniques were used to analyze features such as skull shape (allowing some paleontologists to make claims about sexual dimorphism and social and feeding behavior). Scientific interest in dinosaurs, spurred by spectacular discoveries and new investigative methods, has today grown to such an extent that more scientists currently work on dinosaurs than ever before. People are interested in investigating growth patterns, physiology, and other paleobiological questions, but descriptive and comparative work, and investigations of anatomy and function, are still very much alive. New tools such as CT scanners and 3D software permit an entirely novel approach.

The human side of science

The story of paleontology is not just about discoveries, but also about changes in society and culture, and about people and their ideas and theories. This book shows how key discoveries, or ideas about discoveries, sometimes allowed experts to promote their own views. Although good scientists should report the conclusions of their study honestly and without bias, it is obvious that—particularly during those times when there was great debate about a particularly controversial idea—the conclusions reported by some authorities have been shaped by their own agendas. Examples include Richard Owen's apparent use of dinosaurs during the 1840s as a counter-argument to early views about evolution, and Henry Osborn's use of *Tyrannosaurus*, the largest of the carnivorous dinosaurs, to provide vindication for his views on an "orthogenetic" driving force within evolution.

Entrenched views

It is important to remember that views on dinosaurs—and other aspects of natural history—do not always become established because they are well supported by evidence. Such views are often accepted because they were the first published on the subject, or because nobody ever took the time to challenge them properly. These notions then become uncritically repeated, eventually becoming part of what is known as "textbook orthodoxy." Perhaps due in part to the long period of inertia that afflicted dinosaur research in the mid-20th century, dinosaurs appear to have suffered from this more than many other groups of fossil animal—many views proposed in the early 1900s became entrenched until a new generation of scientists overturned them in the 1960s and 70s. The role of artwork in promoting historical ideas about dinosaurs should not be under-emphasized, and certain views about dinosaurs became enshrined because the artistic reconstructions depicting them were memorable and sometimes wonderful.

Although it is tempting to regard certain old views on dinosaurs as naive, misguided, or even preposterous, we should remember that the people who held these views were often tremendously knowledgeable and experienced, but simply lacked the benefit of hindsight that we are so lucky to have today. Indeed, we should keep in mind that views we consider as undeniably correct today may prove quite wrong in the future.

Above The skills and techniques involved in the discovery and excavation of dinosaurs in the field—keen observation and hard work—have remained largely unchanged over the last 100 years.

Below Advances in technology, such as 3D scanning, have allowed scientists to look anew at the biology and behavior of extinct animals.

What are dinosaurs?

Dinosaurs are one of the most successful and diverse groups of vertebrates in history, dominating the Earth for the better part of 160 million years. They evolved a fantastic array of body types, grew to sizes unparalleled in other land animals, and included the largest terrestrial carnivores that ever lived, as well as some of the most formidable, heavily armored herbivores. Dinosaurs also took to the sky. A compelling body of evidence shows that birds are dinosaurs and the only dinosaur group to survive the mass extinction event at the end of the Cretaceous period.

Skeleton of *Tyrannosaurus rex* All dinosaurs have certain skeletal features in common, and some of the prominent common features are indicated here.

Antorbital fenestra, a large bony opening, is located between the nostril and eye

Neck is long and slender compared to those of members of many other archosaur groups

Shoulder girdle is oriented so that the shoulder socket faces backward and slightly outward

Acetabulum, or hip socket, is fully open, allowing the head of the thigh bone to fit tightly inside

Bony shelf at back of hips provides an extra area for muscle attachment

Head of femur, or thigh bone, is fully offset relative to the bone's shaft

Sacrum—three or more fused vertebrae joined to pelvis—is usually reinforced compared to that of other reptiles

Humerus, or upper arm bone, has a particularly long crest for muscle attachment

Outer fingers (especially the fourth and fifth) are typically reduced in dinosaurs and lost altogether in some groups

Cnemial crest—a bony projection at the top of the tibia—is well developed

Metatarsal bones in feet are usually long and slender; first and fifth metatarsals are often shortened relative to the others

Pubic bone in hips is usually long—a similar length to the thigh bone

Fibula, the outer of the two shin bones, is slender compared to that of most other archosaurs

Ankle is simple and hinge-like; the two largest ankle bones are firmly attached to the bottom of the shin bone, or tibia

The word "dinosaur" is one of the most familiar animal group names: most people can give a rough idea of what sort of creature the word represents. But few can describe exactly what a dinosaur is, and the name often serves as a catch-all for any prehistoric reptile, or indeed any prehistoric animal. Dinosaurs are actually a group of reptiles characterized by a distinct set of anatomical features. In other words, the term "dinosaur" is limited to a select number of organisms that all descended from the same ancestor.

Dinosaurs belong to a group of reptiles called Archosauria, which includes crocodilians and pterosaurs. Unlike other reptiles, archosaurs have a large cavity on the side of the snout called the antorbital fenestra. Within Archosauria, dinosaurs share certain features with pterosaurs and with a group of small Triassic archosaurs regarded as "proto-dinosaurs," such as *Lagerpeton*, *Marasuchus*, and *Silesaurus*. Most proto-dinosaurs were small, lithe, probably carnivorous animals, and dinosaurs have their origins in this body plan.

In earlier books, dinosaurs are described as a group of terrestrial reptiles confined to the Mesozoic Era. This loose definition is not useful, because if birds are included in Dinosauria it follows that dinosaurs are not, in fact, limited to the Mesozoic, nor to a terrestrial existence (penguins and ducks are swimming dinosaurs, for example).

Key dinosaur characters

Dinosaurs are erect-limbed archosaurs in which the head on the thigh bone turns inward completely and fits into a fully open hip socket. The foot bones are long, and the ankle is hinge-like (rather than able to rotate, as is the case in other archosaurs). The outer fingers (the fourth and fifth, but also the third in tyrannosaurids) are typically reduced. Unlike archosaurs, dinosaurs have lost the postfrontal bone in the skull and the deltopectoral crest on the humerus is very long. The shin bone, or tibia, has an enlarged, rectangular lower end. Some dinosaur groups later modified many of these features as they evolved.

Changing views on the Dinosauria

Dinosaurs were first named in 1842, when Richard Owen proposed several features that separate them from all other reptiles. He singled out three main characters: the sacrum containing five fused vertebrae; the large, heavily built limb bones suited for terrestrial movement; and the animals' great size. However, later finds showed that some dinosaurs are small dinosaurs with delicate limb bones.

The concept of Dinosauria as a natural group was challenged during the 1880s, when Harry Seeley noted that it could be divided into two separate groups. Named Ornithischia and Saurischia, these differed in pelvic structure and other details. In the decades that followed, it became generally accepted that ornithischians and saurischians were not each other's closest relatives and that both had arisen from different ancestors.

During the 1970s, palaeontologists noticed that early ornithischians and saurischians shared anatomical features absent in other archosaurs (these included some of the key dinosaur characters listed above). Accordingly, in 1974 Robert Bakker and Peter Galton argued for the resurrection of the clade Dinosauria—a clade is a group of organisms that share a single ancestor. Later studies and discoveries supported the view that dinosaurs form a clade, and additional uniquely dinosaurian characters came to light during the 1980s and 90s.

Above Dinosaurs and pterosaurs are both part of the reptile group Archosauria. Crocodilians and their relatives are also archosaurs, but belong to a fundamentally different branch. This image shows the gigantic Cretaceous alligator *Deinosuchus*.

11

Dinosaur
diversity

Dinosaurs are a huge, diverse group of animals, and in the course of this book we will be looking at species from all the major dinosaurian lineages. Here, we look briefly at dinosaur diversity in order to introduce all the major groups and the names that we use for them.

Compiling the dinosaur family tree The family tree shown here, properly called a phylogeny, has been built up over the last few decades as experts have identified shared characters that unite the different dinosaur groups. The dinosaur phylogeny shown here is generally agreed on, but some details remain controversial and study continues.

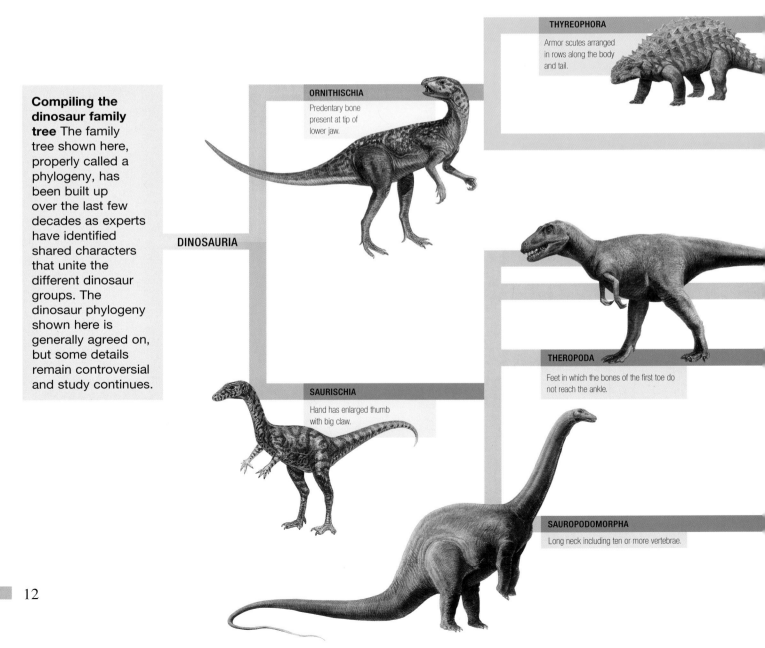

THYREOPHORA

Armor scutes arranged in rows along the body and tail.

ORNITHISCHIA

Predentary bone present at tip of lower jaw.

DINOSAURIA

THEROPODA

Feet in which the bones of the first toe do not reach the ankle.

SAURISCHIA

Hand has enlarged thumb with big claw.

SAUROPODOMORPHA

Long neck including ten or more vertebrae.

Dinosaurs were highly diverse in body shape and also in the ranges of body sizes they evolved. Excluding birds, we know of over 1,400 dinosaur species. However, hundreds are known only from scrappy fossils and a 2008 study found that nearly 52 percent were not supported by good fossil material. About 15 new species are named each year.

Dinosaurs diverged into two major groups early in their evolution: saurischians and ornithischians. Saurischians include the mostly predatory, bipedal theropods and the mostly herbivorous, long-necked sauropodomorphs.

Ornithischians were mainly herbivorous and diverged into one lineage that included the thyreophorans (armored ornithischians, such as stegosaurs and ankylosaurs), and another that included the ornithopods and marginocephalians.

Virtually all paleontologists now regard birds as part of Dinosauria, from which it follows that dinosaurs survived beyond the end of the Cretaceous. Furthermore, the most species-rich part of the dinosaur family tree is represented by the birds. There are approximately 10,000 living bird species, and one estimate suggests that more than a million species lived during the Cenozoic (the last 65 million years).

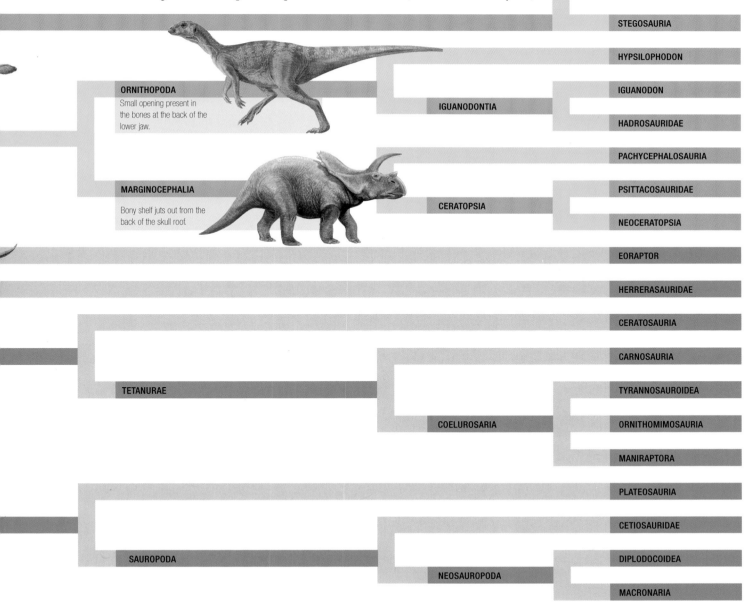

ORNITHOPODA

Small opening present in the bones at the back of the lower jaw.

MARGINOCEPHALIA

Bony shelf juts out from the back of the skull roof.

ANKYLOSAURIA

STEGOSAURIA

HYPSILOPHODON

IGUANODON

IGUANODONTIA

HADROSAURIDAE

PACHYCEPHALOSAURIA

PSITTACOSAURIDAE

CERATOPSIA

NEOCERATOPSIA

EORAPTOR

HERRERASAURIDAE

CERATOSAURIA

CARNOSAURIA

TETANURAE

TYRANNOSAUROIDEA

COELUROSARIA

ORNITHOMIMOSAURIA

MANIRAPTORA

PLATEOSAURIA

CETIOSAURIDAE

SAUROPODA

DIPLODOCOIDEA

NEOSAUROPODA

MACRONARIA

Earth in the time
of the dinosaurs

Dinosaurs were around for a vast span of time, and during this time the Earth underwent some immense changes: the continents split and joined, sea levels rose and fell, and climates changed.

We assume that dinosaurs and other Mesozoic animals were strongly affected by all of these changes, but working out which factors were more important in shaping dinosaurian evolution is controversial. When new possibilities (such as the appearance of a land bridge) presented themselves, dinosaurs probably took advantage and spread to new areas, a type of movement called dispersal. As land masses move, barriers arise that prevent individuals from spreading, causing evolutionary divergence. This process is called vicariance.

Shifting continents and changing coastlines

When dinosaurs first appeared, during the Late Triassic, the continents were united in a single supercontinent, named Pangaea. Accordingly, Late Triassic and Early Jurassic dinosaur diversity was generally similar all around the world. Pangaea began to split up by the end of the Triassic. By the Middle Jurassic, we can see evidence for variation among the dinosaur faunas of Europe, Asia, and North America, and the formation of the Turgai Sea along the western margin of Asia resulted in an isolated Asia. Exceptionally high sea levels during the Jurassic largely submerged Europe and made it a series of archipelagoes.

The North Atlantic had started to open between North America and Europe by the end of the Jurassic, but similar animals in both regions indicate that terrestrial contact was still present as late as the Early Cretaceous. Also during the Late Jurassic, the northern continents diverged from the southern ones, and Pangaea split into Laurasia in the north and Gondwana in the south. The Tethys Sea opened between Africa and Europe, but a reduction in the size of the Turgai Sea during the Early Cretaceous enabled contact between Asia and Europe.

Gondwana became increasingly fragmented during the Cretaceous, initially breaking into a western land mass (South America and Africa) and an eastern one (Indo-Madagascar, Antarctica, and Australasia). South America and Africa parted during the Early or mid Cretaceous, and a united India and Madagascar moved northeast from Antarctica. By the end of the Late Cretaceous, Madagascar and India had parted ways and India would eventually collide with

Below This timeline shows the major geological segments of the past 540 million years and the evolutionary changes associated with each segment. An era is a large segment of time that is divided into smaller segments called periods. Dinosaurs are associated mainly with the Mesozoic Era. After evolving in the Triassic Period, they dominated terrestrial life in the Jurassic and Cretaceous. At the end of the Mesozoic, all dinosaurs except birds became extinct. Birds thrived in the Cenozoic and continue to do so today, continuing the important global presence of dinosaurs that has so far lasted more than 200 million years.

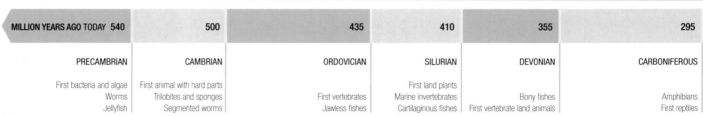

MILLION YEARS AGO TODAY 540	500	435	410	355	295
PRECAMBRIAN	CAMBRIAN	ORDOVICIAN	SILURIAN	DEVONIAN	CARBONIFEROUS
First bacteria and algae	First animal with hard parts		First land plants		
Worms	Trilobites and sponges	First vertebrates	Marine invertebrates	Bony fishes	Amphibians
Jellyfish	Segmented worms	Jawless fishes	Cartilaginous fishes	First vertebrate land animals	First reptiles

continental Asia. Toward the end of the Late Cretaceous, a rift that would form the Tasman Sea began to appear along the eastern margin of eastern Gondwana, and a land mass that today includes New Zealand and New Caledonia broke away and moved rapidly eastward. Australia moved northeast at the same time.

By the Late Cretaceous, Europe was home to endemic, island-dwelling dinosaurs as well as to Gondwanan kinds that had presumably invaded from Africa. Split in two by the Western Interior Seaway, eastern and western North America may have had quite different faunas. Indeed, the Late Cretaceous dinosaurs of western North America have strong affinities with the dinosaurs of eastern Asia, whereas those of eastern North America are altogether different. Although we are familiar with the Late Cretaceous dinosaurs of western North America, which include tyrannosaurids and horned dinosaurs, more archaic dinosaurs were the norm on the southern continents.

Mesozoic climates

The Mesozoic Earth was in what is known as a greenhouse phase, when warm, equable climates existed worldwide and ice caps were absent or at best small. Huge deserts covered much of Pangaea, with the relatively small coastline and low sea levels exaggerating the dry, strongly seasonal climate. During the Jurassic, the break-up of Pangaea and an increasing build-up of seafloor crust (caused by the emergence of lava at active continental margins) produced higher sea levels and cooler, more humid conditions. The Cretaceous world was also warm, with lush vegetation close to the poles.

Mesozoic climates remain a topic of much debate, and why the Mesozoic world was so warm has not been adequately explained. During the Cretaceous, continental interiors may have been cool, and glaciers and continental ice may have covered some regions during the Jurassic and Cretaceous.

Below The typical environments of the Triassic, Jurassic, and Cretaceous worlds would have appeared very different due to key climatic and biological changes during each period. The Triassic world was dominated by deserts, although forests of cycads and other plants were widespread. During the Jurassic, conifers, cycads, and ferns flourished in the more humid conditions. By the Cretaceous, modern types of plants had evolved and some environments would have had a very "modern" appearance.

TRIASSIC

JURASSIC

CRETACEOUS

During the Triassic, the giant super-continent Pangaea stretched from northernmost regions to southernmost. It was surrounded by the Panthalassic Ocean.

Toward the end of the Jurassic, Pangaea split in two to form Laurasia in the north and Gondwana in the south, with the Tethys Sea between.

The Atlantic opened during the Cretaceous, and Gondwana split, forming several fragments. Africa moved north, closing the Tethys Sea.

During the Cenozoic, Africa and India collided with Eurasia. The Atlantic widened and a bridge formed between the Americas.

PALEOZOIC ERA		MESOZOIC ERA	CENOZOIC ERA
248	203	144	65	1.75	TODAY
PERMIAN	TRIASSIC	JURASSIC	CRETACEOUS	TERTIARY
	First dinosaurs			Dominance of mammals
	Turtles			
	Lizards, and crocodiles	Dinosaurs dominate		QUATERNARY
Mammal-like reptiles	First mammals	First birds	Decline of dinosaurs	Modern humans

Pioneering Dinosaur Discoveries

The earliest discoveries of fossils that came to be recognized as dinosaur bones took place in the early 19th century in western Europe, but these early finds were scrappy and difficult to interpret— for years they were thought to be from gigantic, lizard-like animals, and it was decades before a closer idea of the true appearance of dinosaurs was revealed. North American dinosaurs were first reported in 1850 and became increasingly important. By the late 1800s, spectacular dinosaur discoveries were taking place in both Europe and the United States.

Left In this classic early illustration, *Megalosaurus* and *Iguanodon* are both shown as short-necked quadrupeds. Early reconstructions of dinosaurs were radically inaccurate by modern standards because so few bones had been discovered—dinosaurs were thought to resemble scaled-up versions of modern reptiles.

The 19th century

ALASKA

In 1842, the trailblazing British scientist Richard Owen announced the discovery of the dinosaurs to great acclaim. He described them as immense animals with thick limb bones and strong, reinforced hips. They were the "most perfect modifications of the Reptilian type," and "must have played the most conspicuous parts, in their respective characters as devourers of animals and feeders upon vegetables, that this world has ever witnessed in oviparous and cold-blooded creatures."

Two decades earlier, scientists already knew of the existence of gigantic sea reptiles in the distant past, but by the time of Owen's publication, clues were emerging that equally impressive reptiles had lived on land as well. The first of these "dinosaurs-to-be" was a huge carnivore, discovered in an English rock known as the Stonesfield Slate, and published by William Buckland in 1824 under the name *Megalosaurus*. Buckland imagined it to be a fearsome predatory lizard with a total body length in the region of 60–70 ft (18–21 m).

A second "dinosaur-to-be" was found by Gideon Mantell, a medical doctor and amateur fossil collector from the small market town of Lewes in East Sussex, in southern England. Mantell built up an impressive collection of fossils from his neighborhood over the years, first publishing on them in 1822. However, the fact that he lived outside the socio-political center of 19th-century England counted against him, despite his best efforts to make his way in London scientific circles. By the early 1820s, Mantell had obtained several unusual fossil teeth that came from the Tilgate Grit (now known to be Lower Cretaceous in age), and in 1825 he named *Iguanodon* for these remains. Since the teeth resembled those of iguanas, he regarded *Iguanodon* as a lizard-like animal but enormous in size, its length perhaps exceeding 100 ft (30 m).

A third "dinosaur-to-be" was, like *Iguanodon*, also from the county of East Sussex. The specimen was uncovered by workmen and blasted into fragments that Mantell later managed to fit together, and in 1833 he named it *Hylaeosaurus*. It is a big, armored reptile with large, conical spines arranged on the neck and shoulders, and has remained poorly known— certainly it has never become as familiar to the public as *Megalosaurus* or *Iguanodon*.

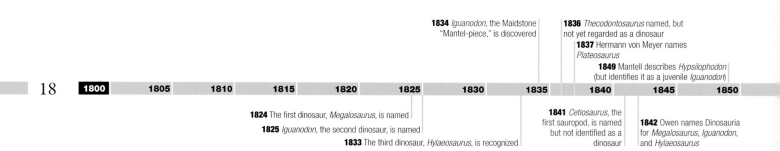

1834 *Iguanodon*, the Maidstone "Mantel-piece," is discovered

1836 *Thecodontosaurus* named, but not yet regarded as a dinosaur

1837 Hermann von Meyer names *Plateosaurus*

1849 Mantell describes *Hypsilophodon* (but identifies it as a juvenile *Iguanodon*)

18   **1800**   1805   1810   1815   1820   1825   1830   1835   **1840**   **1845**   **1850**

1824 The first dinosaur, *Megalosaurus*, is named

1825 *Iguanodon*, the second dinosaur, is named

1833 The third dinosaur, *Hylaeosaurus*, is recognized

1841 *Cetiosaurus*, the first sauropod, is named but not identified as a dinosaur

1842 Owen names Dinosauria for *Megalosaurus*, *Iguanodon*, and *Hylaeosaurus*

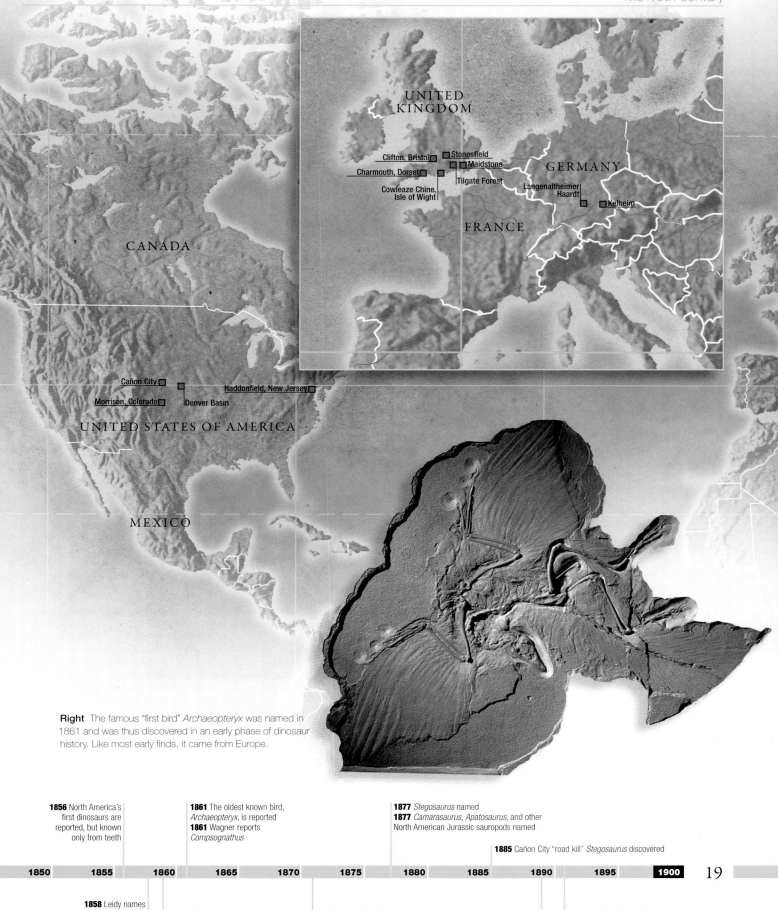

Right The famous "first bird" *Archaeopteryx* was named in 1861 and was thus discovered in an early phase of dinosaur history. Like most early finds, it came from Europe.

1856 North America's first dinosaurs are reported, but known only from teeth

1861 The oldest known bird, *Archaeopteryx*, is reported
1861 Wagner reports *Compsognathus*

1877 *Stegosaurus* named
1877 *Camarasaurus*, *Apatosaurus*, and other North American Jurassic sauropods named

1885 Cañon City "road kill" *Stegosaurus* discovered

| 1850 | 1855 | 1860 | 1865 | 1870 | 1875 | 1880 | 1885 | 1890 | 1895 | **1900** | 19 |

1858 Leidy names *Hadrosaurus* from New Jersey

1859 Early armored dinosaur *Scelidosaurus* is named

1871 Phillips describes good *Cetiosaurus* specimen

1889 *Triceratops* named

1891 Marsh produces first *Stegosaurus* reconstruction

Above William Buckland, shown here delivering a lecture, is best known for having described *Megalosaurus*. He was a polymath, a brilliant thinker, and also a member of the clergy.

Owen's Dinosauria

Megalosaurus, *Iguanodon*, and *Hylaeosaurus* were all remarkable discoveries, but these creatures were not regarded as close relatives until 1842, when Owen proposed that they be united in a new group, which he called the Dinosauria. He argued that dinosaurs—the name means "fearfully great lizards"—resembled large modern land mammals such as elephants and rhinos in their terrestrial habits, reinforced hip regions, and massive, elephantine limbs. However, it was simply a coincidence that the three species known at this point shared a special reinforced pelvic region: more recent discoveries have shown that this feature is not a defining character of dinosaurs as once was thought. Interestingly, despite emphasizing their size, Owen also inadvertently "downsized" the dinosaurs. For example, he suggested that *Megalosaurus* and *Iguanodon* were both about 30 ft (9 m) long.

Moreover, discoveries made during the 1860s showed that dinosaurs were not all elephantine monsters, but also included small species. The bird-like anatomy of the small dinosaurs *Hypsilophodon* (from England) and *Compsognathus* (from Germany) enabled Thomas Huxley—"Darwin's bulldog"—to propose an affinity between dinosaurs and birds. *Archaeopteryx*, named from Germany in 1861, demonstrated that birds themselves lived alongside dinosaurs.

Owen's views on dinosaurs became more familiar to the public than did those of his colleagues Buckland and Mantell because life-sized models of "Owenian dinosaurs" were created for a special exhibition of extinct animals—known as the Dinosaur Court—opened in the grounds of the Crystal Palace, London, in 1854. The Crystal Palace Company had initially approached Mantell, but he had turned them down.

Global dimensions

Dinosaur fossils in North America were first reported during the 1850s. In 1855, American geologist Ferdinand Vandiveer Hayden collected fossil teeth from the Cretaceous badlands of what is now Montana. On his return to Philadelphia in 1856 he passed them to paleontologist Joseph Leidy. Although dinosaur tracks were discovered in the Connecticut Valley decades earlier, and described by the Reverend Edward Hitchcock in 1836, they were thought at the time to be bird tracks. The teeth collected by Hayden are therefore seen as the first North American dinosaur remains (*see* pp.32–33). They indicated that dinosaurs with similarities to *Iguanodon* and *Megalosaurus* awaited discovery on the continent. This was confirmed in 1858, when *Hadrosaurus*—the first duckbilled dinosaur to be found—was discovered in New Jersey. It was clearly related to *Iguanodon*, yet its bones suggested a kangaroo-like body.

Iguanodon was itself to be re-imagined entirely, due to the discovery of several specimens at Bernissart in Belgium in 1878. Having studied the new skeletons, Belgian paleontologist Louis Dollo showed that *Iguanodon* was shaped more like a giant kangaroo than the quadrupedal, lizard- or rhino-like animal originally pictured by Mantell and Owen.

By the end of the century, dinosaurs had become reasonably well known as a fascinating and diverse group of animals. In North America, stegosaurs were discovered in the 1870s and the giant horned dinosaur *Triceratops* was named during the 1880s. Sauropods, which had first been discovered during the 1840s in England but had remained enigmatic and mysterious, became better understood, thanks to the naming of *Camarasaurus* from Colorado, in 1877. This was followed soon after by many other Late Jurassic relatives. It was clear that a golden age of discovery had begun.

Above Richard Owen's elephantine dinosaurs were immortalized as life-sized models displayed to this day at Crystal Palace, London. Before the models were put on display in 1854, a dinner was held inside the *Iguanodon*.

21

Megalosaurus, the first theropod

In 1824, William Buckland gave the name *Megalosaurus* to a collection of giant reptile bones that came from the Stonefield Slate of Oxfordshire, England. Though not associated with other fossil reptiles or named as a dinosaur until 1842, this was the very first dinosaur to be described.

The story of what was eventually to become *Megalosaurus* actually began long before 1824. Various bones from Stonesfield were sent to the Ashmolean Museum in Oxford in the late 1700s, and in 1818 they were examined by George Cuvier— the great French pioneer of comparative anatomy— while he was Buckland's guest (Buckland was then the Ashmolean's Director). Cuvier and Buckland concluded that the Stonesfield animal was an immense, lizard-like reptile, about 40 ft (12 m) long and as bulky as an elephant. The creature was likened to the living monitor lizards and became known as the "Stonesfield monitor."

DISCOVERY PROFILE

Name	*Megalosaurus bucklandii*
Discovered	Stonesfield, Oxfordshire, England, by anonymous quarrymen between late 1700s and early 1800s
Described	By William Buckland, 1824
Importance	The first dinosaur to be discovered, and first theropod to be discovered
Classification	Saurischia, Theropoda, Carnosauria, Spinosauroidea

The race to publish

Some think it likely that Buckland's publication of *Megalosaurus* was prompted by news that Gideon Mantell was soon to publish *Iguanodon*. The latter was a genuinely new discovery, the Oxford megalosaur was not: Buckland had known about it since before 1818, and some Stonefield bones were collected as early as the late 1700s. The best known *Megalosaurus* bone—an incomplete dentary with several teeth in place—was acquired by the Oxford Anatomy School at Christchurch College in 1797. In 1822 James Parkinson published the name *Megalosaurus* together with an illustration of one of the teeth.

Right *Megalosaurus* was originally assumed to be quadrupedal. This reconstruction from 1854 shows one early rendition of this dinosaur.

In 1824 Buckland finally published the specimen, thereby producing the first scientific description of a dinosaur. His account was brief, focusing on the dentary, some vertebrae (including a sacrum), rib fragments, and pelvic bones. He noted that the bones must have belonged to "several individuals of various ages and sizes," and speculated that *Megalosaurus* might have been amphibious. *Megalosaurus* was without an official species name, however, until 1827, when Mantell named it *Megalosaurus bucklandii*.

The quintessential theropod

Any suggestion that a giant predatory lizard once roamed the English countryside must have been shocking to the general public. However, within the space of a few years, people really were imagining *Megalosaurus* to be immense and lizard-like. A reconstruction that appeared in Penny Magazine in 1833 certainly depicts it in this manner, although it gave the beast rather short legs, despite the fact that Buckland had illustrated long hind limb bones.

Ideas about the appearance of large predatory dinosaurs changed markedly as more complete discoveries were made in continental Europe and North America. Nevertheless, as the very first predatory dinosaur to be named, *Megalosaurus* became the quintessential theropod, and for decades theropod remains from around the world, and from rocks of varying geological age, were all labeled as *Megalosaurus*.

Above Like all predatory dinosaurs, *Megalosaurus* was in fact a bipedal predator that used its hands in tackling prey. Here, a megalosaur attacks an unlucky herbivore.

CONFLICTING THEORIES

Buckland was not only a geologist and paleontologist, but also a member of the clergy (he became Dean of Westminster in 1845), and he tried to reconcile geological evidence with the Biblical story of creation, publishing a book on this subject in 1820. It might have been difficult for Buckland to get *Megalosaurus* to fit into his revised view of Biblical creation, and some experts have suggested that this problem might explain why he delayed publishing a description of the dinosaur. Later, Buckland argued that the Bible could be reinterpreted in a manner that permitted the new evidence from geology and paleontology to fit into the story of Genesis. He even claimed that the formidable teeth, claws, and size of *Megalosaurus* demonstrated the existence of a creator who favored the design of carnivores that were able to end the lives of their victims with swift mercy. Ideas such as these put Buckland at loggerheads

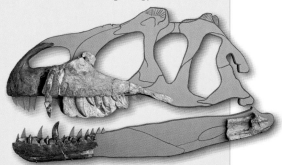

Above Discoveries made in the U.S.A. and elsewhere have revealed that *Megalosaurus* had a skull shaped something like this.

with Biblical literalists, who countered that *Megalosaurus* must have been herbivorous given that carnivory and death did not appear until Adam and Eve sinned.

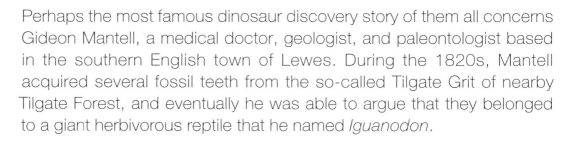

Mantell and the iguanodonts

Perhaps the most famous dinosaur discovery story of them all concerns Gideon Mantell, a medical doctor, geologist, and paleontologist based in the southern English town of Lewes. During the 1820s, Mantell acquired several fossil teeth from the so-called Tilgate Grit of nearby Tilgate Forest, and eventually he was able to argue that they belonged to a giant herbivorous reptile that he named *Iguanodon*.

Above British scientist Gideon Mantell is best remembered today for his discovery of *Iguanodon*, the second dinosaur to be named. However, he also worked on ancient mollusks, plants, and many other fossils.

Exactly how these specimens were uncovered is shrouded in mystery. Mantell's wife Mary is often credited with their discovery, and Mantell wrote on several occasions that she did make the initial finds. However, it has also been argued that neither of them discovered these teeth and that they were instead bought from quarrymen.

After consulting Cuvier and other experts, Mantell grew confident that the teeth were reptilian, and this was confirmed in 1824 when he was able to examine the teeth of a modern iguana. Mantell's investigations had initially been complicated by the erroneous belief that the Tilgate Grit was "alluvial" and geologically young, rather than Mesozoic in age. Today, the Tilgate Grit is regarded as part of the Lower Cretaceous Wealden Supergroup.

Satisfied that the Tilgate teeth were from a huge iguana-like herbivore, Mantell announced his new animal, by then named *Iguanodon*, to the Geological Society in London in 1825. Assuming *Iguanodon* to be iguana-like in shape, he proposed that it was more than 60 ft (18 m) long. Mantell earned much recognition from this discovery: his paper on *Iguanodon* was also read before the highly prestigious Royal Society of London, and in that year he was elected to both the Royal Society and the Geological Society.

Picking up the pieces

Mantell had no real idea of what *Iguanodon* might have looked like until 1834, when a disarticulated partial skeleton was discovered at a quarry in Maidstone, Kent, in southeast England. The quarry owner, William Harding Bensted, sold the specimen to Mantell in July of that year. Bensted was himself an amateur fossil collector who regularly corresponded with Mantell, Owen, and other paleontologists and geologists, and he was later to have several species named after him.

Realizing that the Maidstone specimen was of great interest, Bensted gathered up the numerous fragments that had been scattered during the gunpowder blast that unearthed the specimen. One large section had already been loaded aboard a barge destined for London, but Bensted was able to prevent this loss and reunite the pieces of the specimen. He constructed a shed over the remains and set about removing some of the surrounding rock in order to better

---

DISCOVERY PROFILE

Name	*Iguanodon*
Discovered	Tilgate Forest, Sussex, England, by Gideon and Mary Mantell (or by anonymous quarrymen), early 1820s
Described	By Gideon Mantell, 1825
Importance	The first plant-eating dinosaur and the first ornithischian to be discovered
Classification	Ornithischia, Ornithopoda, Iguanodontia

reveal the bones, a task that occupied him for a month. With the help of geologist George Ramsgate, Bensted identified the specimen as *Iguanodon*, and this was confirmed by Mantell on his visit in June 1834.

The Maidstone "Mantel-piece"

Thanks to financial assistance provided by two friends, Mantell obtained the specimen and quickly set to work preparing it further. It combined teeth similar to the Tilgate specimens with a jumbled assortment of limb bones, ribs, and vertebrae. In keeping with Mantell's earlier idea that *Iguanodon* was lizard-like, he concluded that the animal was more than 100 ft (30 m) long, but bulkier and perhaps different in proportions from modern lizards.

In a sketch produced by Mantell—and never intended for publication—bones now known to be from the pelvis were misidentified as collarbones. He also provided the creature with a conical nose horn, an object that later turned out to be a spike-like thumb bone. The "Mantel-piece", as it became known, did not preserve

such a "horn", but one had been discovered at Tilgate and described by Mantell in 1827. This early blueprint for *Iguanodon* was soon adopted by artists: for example, in 1838 John Martin illustrated *Iguanodon* in this guise for Mantell's book *Wonders of Geology*. Mantell modified this view in later years, noting in 1851 that the relatively short forelimbs might have been used to grab and pull down foliage, rather than simply to support weight.

Obtained from a rock known as Kentish Rag, the Maidstone "Mantel-piece" was substantially younger than the Tilgate Grit remains and probably represented a substantially different sort of iguanodontian dinosaur. With the benefit of modern knowledge of iguanodontian anatomy, David Norman was able to show in the 1990s that the "Mantel-piece" preserves the better part of the animal's body, including most of the bones of the limbs and limb girdles.

Above Shown here battling with a *Megalosaurus*, *Iguanodon* was thought by Mantell and others to be lizard-like, with a horn on its nose.

Below Today we know that *Iguanodon* was not the short-necked, horned, lizard-like animal imagined by Mantell. The spike belonged on the thumb, not the nose.

An early dinosaur from Bristol

Most of the early dinosaur discoveries—such as *Megalosaurus*, *Iguanodon*, and *Hadrosaurus* for example—are well known. Stories about their discoveries are related in most of the dinosaur literature. But one of the very earliest dinosaurs to be described is seldom discussed and all but unknown outside the field of dinosaur research.

DISCOVERY PROFILE

Name	*Thecodontosaurus antiquus*
Discovered	Clifton, Bristol, England, by Henry Riley and Samuel Stutchbury, 1834
Described	By Henry Riley and Samuel Stutchbury, 1836
Importance	The first Triassic dinosaur to be discovered, and the first sauropodomorph to be discovered
Classification	Saurischia, Sauropodomorpha

Below
Thecodontosaurus was a small, delicately built quadruped, and very different from the gigantic sauropods that evolved from its close relatives.

The creature in question is *Thecodontosaurus*, which in 1836 became the fourth British dinosaur to be given a name. Since this followed the naming of the poorly known taxa *Streptospondylus* (in 1832) and *Macrodontophion* (in 1834), *Thecodontosaurus* was only the sixth dinosaur ever to be described. This dinosaur was not a visually impressive animal, but it has been important in developing our understanding of dinosaur evolution.

The socket-toothed lizard

Thecodontosaurus was discovered in 1834 after the surgeon and teacher Henry Riley and Samuel Stutchbury, later of the Royal College of Surgeons, began collecting reptile fossils from the limestone quarries of Clifton, Bristol, in southwest England. Their original material consisted of an incomplete lower jaw, vertebrae, ribs, and several limb bones. Numerous other bones were removed from the same site in the years that followed. In 1836, Riley and Stutchbury published an article announcing the discovery of this new fossil animal: it was clearly a reptile, and the presence of serrated, leaf-shaped teeth set within sockets led them to name it *Thecodontosaurus* (meaning "socket-toothed lizard"). It was also small: the lower jaw was less than 3 in (7.5 cm) long. In 1842, Richard Owen proposed that it should be included among a group of reptiles he called the "thecodonts", and in 1859 he formally

THECODONTOSAURUS RECONSTRUCTED

Thecodontosaurus was known only from disarticulated material until 1952, when what was thought to be a juvenile of this taxon was discovered at Pant-y-ffynnon Quarry in South Glamorgan, Wales. Recently, this find has been shown to represent a distinct animal, dubbed *Pantydraco caducus*. *Pantydraco* was found at the bottom of what was originally a deep fissure between two blocks of limestone. The unlucky individual seems to have fallen into this fissure, become trapped, and eventually starved to death.

named the group Thecodontia. Later the group was placed within the early archosaurs.

The true, dinosaurian affinities of *Thecodontosaurus* were missed partly because the remains were so small, but also because they lacked the long, fused sacrum and robust limb bones that Owen regarded as characteristic of this group. Owen noted that *Thecodontosaurus* resembled monitor lizards except for the distinct presence of tooth sockets. However, in passing he also mentioned a similarity with *Megalosaurus*. It seems that Owen and his contemporaries imagined *Thecodontosaurus* to look essentially like a reasonably large lizard, albeit one with unusual teeth, and *Thecodontosaurus* remained an enigma for some time afterward.

Thecodontosaurus and the Dinosauria

In 1870, Thomas Huxley noted that the hip and hind limb bones of *Thecodontosaurus* were more similar to those of *Megalosaurus* than they were to those of any other fossil reptile. He therefore argued that *Thecodontosaurus* should be included within the Dinosauria. Because the teeth of *Thecodontosaurus* were thought by Huxley to most resemble those of *Hylaeosaurus* and other armored dinosaurs, he included the Bristol dinosaur within the Scelidosauridae. Huxley's "Scelidosauridae" contained not only *Scelidosaurus* (*see* pp.36–37), but also dinosaurs today known to be ankylosaurs.

During the 1890s, *Anchisaurus* and other new "Triassic" dinosaurs from North America

(they were actually from the Early Jurassic), led Othniel Marsh to note a similarity between *Thecodontosaurus* and these forms, and he united them in a family thought to belong among the theropods. It was later realized that *Thecodontosaurus*, *Anchisaurus*, and their relatives were close to the ancestry of the immense sauropods, and in 1932 all of these dinosaurs were brought together in a group called the sauropodomorphs. *Thecodontosaurus* was imagined as a small, prototype sauropodomorph, and its great importance lay in the fact that it was small and Triassic in age. Indeed, it was the very first Triassic dinosaur ever reported.

A narrow escape

In 1940, an air raid during World War II destroyed the geology gallery of Bristol City Museum, where the *Thecodontosaurus* specimens were kept. Some of the material survived and today bears scorch marks, having been recovered from the burning museum wreckage. Fortunately, much additional material had been sent elsewhere long before the war: some resides today at the Natural History Museum; some was given to the Academy of Natural Sciences in Philadelphia; and yet more was presented to Othniel Marsh for the Peabody Museum of Natural History at Yale University. In all, more than 240 of the original specimens remain, and a survey published in 2000 showed that much of it corresponds to the original material figured by Riley and Stutchbury.

Above Until recently, it was generally thought that most, or all, of the original *Thecodontosaurus* material had been destroyed during World War II. We now know that many bones survived, including this incomplete lower jaw.

27

Fossil hunting

Despite technological advances, the practice of finding dinosaur fossils has not changed substantially over time. Fossils are still discovered as they always have been, by people exploring an area and scouring the ground with their eyes. Virtually all new dinosaurs and other fossils have been discovered by chance, typically when paleontologists or fossil collectors explore a region that has yielded fossils in the past. Keen observational skills count above all else.

Most fossils come from places where erosion is active and sediment is constantly being broken away. Cliffs, badlands, and windy deserts therefore provide most of the best dinosaur-bearing sites, and this is why new dinosaurs mostly come from remote regions. However, people's digging and quarrying activities also expose new sediment layers, and many dinosaurs have been discovered in quarries and other excavated sites.

Right Fossils that erode from the sides of cliffs, such as this one in Dinosaur Provincial Park, Canada, may be visible from a long way off. Fossil hunters pay a lot to attention to cliff-faces and to the debris that collects beneath.

Below left Detailed records of locality data must be kept when fossils are found. The standard tool is the notebook, but GPS recorders are increasingly used as well.

Below right Fossil footprints are usually visible only when exposed on the surface, and once there they may be subject to rapid erosion.

The discovery of sauropods

Today, sauropods are very familiar to us, their often huge body size, long necks, and column-like limbs combining to form a highly distinctive body shape. Our view of sauropods comes mainly from key finds that occurred in the United States during the 1870s and 80s—but the interpretation of earlier sauropods remains is an interesting story.

Numerous details of sauropod anatomy, the environments in which their remains were preserved, and other details, show that sauropods were land-living animals; modern reconstructions routinely depict sauropods as fully terrestrial. For much of the 20th century, however, sauropods were thought to be aquatic or amphibious animals that lived in swamps and lakes. Quite why this "aquatic hypothesis" became entrenched in both the scientific and popular literature is an intriguing question, and it seems to owe its origin to the earliest, 19th-century interpretations of these fascinating dinosaurs.

Owen's aquatic cetiosaurs

Sauropod bones were first described in 1841, although at this time the name sauropod was not in use (it was not coined until 1878). In 1841, Richard Owen named *Cetiosaurus*, meaning "whale lizard," for various remains—gigantic vertebrae and limb bones—that had come from Oxfordshire and elsewhere in England. Owen did not recognize the true, dinosaurian identity of these bones and simply remarked that they belonged to a giant predator that must have preyed on crocodilians and plesiosaurs.

Owen never provided a reconstruction of his cetiosaurs, but we do know that he imagined them to be aquatic, based on similarities between their vertebrae and those of whales (*see* box, facing page). He also appears to have thought that cetiosaurs resemble modern crocodilians in shape. We do know that, based on his interpretation of *Cetiosaurus* tail vertebrae, he believed cetiosaurs to have deep, laterally compressed tails like those of crocodiles. He also compared the bones of the shoulder and pelvic girdles to those of plesiosaurs and argued that the limb bones are most like those of crocodilians, and he knew that strong claws are present on the

DISCOVERY PROFILE

Name	*Cetiosaurus*
Discovered	Chipping Norton, Oxfordshire, England, by Mr. Kingdon, *c.*1841, with additional bones from elsewhere being discovered by anonymous quarrymen
Described	By Richard Owen, 1841
Importance	The first sauropod to be discovered
Classification	Saurischia, Sauropodomorpha, Sauropoda

Left The first good-quality *Cetiosaurus* skeleton was found in 1868. It included many bones, such as this shoulder blade, that provided information on the shape of this dinosaur.

WHALE-LIKE CROCODILIAN

The spongy internal texture of *Cetiosaurus* vertebrae reminded Owen of that seen in the similarly huge vertebrae of whales. He noted that the shape of the vertebrae and the coarse internal texture of the limb bones are also whale-like. As a result, he thought some cetiosaurs rivaled modern whales in size and saw them as "strictly aquatic and most probably of marine habits."

Left Most early sauropod finds were large, isolated vertebrae such as this. This specimen is from the Jurassic of eastern England.

toes. He had no idea about the skull anatomy of these animals, nor was it clear at the time that *Cetiosaurus* actually has—as is typical of sauropods—a remarkably long neck.

Owen decided to place *Cetiosaurus* within Crocodilia, and this explains why cetiosaurs were not included in his 1842 recognition of the Dinosauria. Other scientists, however, argued for the dinosaurian identity of these animals and by the 1870s, Owen agreed with them. His identification of cetiosaurs as giant marine crocodiles has mostly been forgotten about today, but it may have been far more influential than is generally realized: Owen's conclusion that cetiosaurs were "strictly aquatic" seems to have been adopted by others, and indeed many scientists in the 19th and 20th centuries wrongly assumed aquatic habits for these creatures.

A taxonomic wastebasket

In his original report on *Cetiosaurus*, Owen did not coin any new species names, but in 1842 he named six, with a further seven being added later by other authors. Given that *Cetiosaurus* was the only named sauropod at this time, it stands to reason that any sauropod remains were placed in this genus. Unfortunately, *Cetiosaurus* has become a so-called "taxonomic wastebasket" for assorted diverse Jurassic and Cretaceous sauropods, and not until 2003 did sauropod experts resolve this muddle. Some British sauropods placed in *Cetiosaurus* can now be shown to be diplodocoids or brachiosaurs; others are indeterminate and cannot be identified more precisely than "sauropod."

However, one animal has been associated with the name *Cetiosaurus* most consistently: the Middle Jurassic English species *C. oxoniensis*.

This species has been shown in recent studies to be a relatively primitive sauropod that can reliably be distinguished from other sauropods on the basis of diagnostic features in its vertebrae, chevrons, and hip bones. Two reasonably good specimens of this species are now known: one was discovered at Bletchingdon, Oxfordshire, England, in 1868 and described by John Phillips of Oxford University in 1871; and the other was found at Great Casterton, Rutland, England, in 1968 and fully described in 2002.

31

The first duckbilled dinosaur

Until the 1850s, the discovery of dinosaurs was an entirely European affair. But in the decades leading up to the turn of the century, North American dinosaurs took center stage. This sea change began in the 1850s, when fragmentary fossils from what is now Montana were sent to Joseph Leidy at the Academy of Natural Sciences in Philadelphia.

Above This modern reconstruction of a hadrosaur skeleton correctly shows a horizontal body and tail and a body that can be posed both bipedally and quadrupedally.

A skilled anatomist, Leidy was also an authority on mineralogy, paleontology, and parasitology, and he also had a great interest in botany and zoology. The new fossils were sent by geologist Ferdinand Vandiveer Hayden, and Leidy named them *Palaeoscincus*, *Trachodon*, *Troodon*, and *Deinodon*; all of these were based on isolated teeth alone. Although *Troodon* and *Palaeoscincus* were thought to belong to lizards, *Trachodon* was noted by Leidy as being similar to *Iguanodon*, and *Deinodon* clearly resembled *Megalosaurus*. These were the first named American dinosaurs, but their fossils were neither impressive nor particularly informative.

The bulky lizard

Far more impressive was the 1858 discovery by William Parker Foulke of a great many bones and teeth at Haddonfield, New Jersey. Part of a jaw, hind and forelimb bones, numerous vertebrae, and pelvic bones were recovered, all described as ebony black and very heavy. Foulke was alerted to the discovery of fossil bones at the Haddonfield site by his neighbor John E. Hopkins, who had discovered numerous bones 20 years before but allowed visitors to carry them off. Leidy was soon informed of this discovery and realized that these remains represented another *Iguanodon*-like herbivore, which he named *Hadrosaurus foulkii*. *Hadrosaurus* means "bulky lizard" and was a clear reference to the animal's size.

With only a fragment of lower jaw to go on, Leidy assumed (as Mantell had for *Iguanodon*) that *Hadrosaurus* has an iguana-like skull. Assuming the number of trunk vertebrae to be the same as for living crocodiles and iguanas, and the tail to have 50 vertebrae, he proposed that the animal was about 25 ft (7.5 m) in length.

Bipedal or quadrupedal?

Leidy noted that the hind and forelimb bones of *Hadrosaurus* have quite different proportions, and deduced that it stood in an erect, bipedal posture. He suggested that it may have behaved like a gigantic kangaroo, leaning back on its tail and feeding from high up in the foliage.

In 1851, Mantell noted that *Iguanodon* may have used its forelimbs to manipulate plants while adopting a bipedal pose, so Leidy was not the first to argue for bipedality in a dinosaur. However, Leidy also remarked that *Hadrosaurus* probably walked quadrupedally and he also suggested that it was amphibious. This idea of

DISCOVERY PROFILE

Name	*Hadrosaurus foulkii*
Discovered	Haddonfield, New Jersey, U.S.A., by William Parker Foulke, 1858
Described	By Joseph Leidy, 1858
Importance	The first North American dinosaur and first hadrosaur skeleton to be discovered
Classification	Ornithischia, Ornithopoda, Iguanodontia, Hadrosauridae

amphibious habits remained associated with hadrosaurs for a long time afterward, even though many details of their anatomy contradicted such a lifestyle.

America's first mounted dinosaur

We have a detailed understanding of Leidy's view of this dinosaur thanks to Englishman Waterhouse Hawkins's reconstruction of its skeleton. Using plaster replicas in place of the bones that were unknown, this was displayed in Philadelphia in 1868, and replicas were sent to Washington, D.C., New York, and Chicago. Arranged in the kangaroo-like posture favored by Leidy, the skeleton soon earned the nickname "kangaroo lizard." It was the first mounted dinosaur skeleton to be exhibited in the U.S., and was a remarkable achievement.

In 1865 Leidy argued that the teeth of *Hadrosaurus* are arranged in close apposition, despite the fact that only a fragment of its jaw and a few isolated teeth were preserved. But his view was confirmed by the later discovery of complete hadrosaur tooth batteries. Leidy's view of the appearance and biology of *Hadrosaurus* paved the way for further work and was clearly adopted by Edward Cope. In 1868, Cope described *Hadrosaurus* in a similar vein to Leidy, noting that it was amphibious, habitually used its forelimbs to grab plant food, and walked on its hind limbs alone, at least at times.

Cope modified Leidy's view of the hadrosaur in 1883, when he described the first complete hadrosaur skull. Rather than short, iguana-like skulls, these dinosaurs have remarkable long jaws expanded at their toothless tips into a broad, duck-like bill. More complete hadrosaur discoveries, described during the 1890s by Cope's rival Othniel Marsh, rendered Leidy's original model increasingly obsolete.

Below Like all hadrosaurs, *Hadrosaurus* had a long skull with a broad, duck-like beak.

Archaeopteryx
arrives

Perhaps the most famous extinct organism in the world is *Archaeopteryx*, the German *urvogel* or "first bird." At the time of writing, there are 10 known specimens of this Late Jurassic feathered animal, although it is generally agreed that not all belong to the same species. The first of them was discovered at an early, pivotal stage in our understanding of dinosaur evolution.

Known today as the "London specimen," the first *Archaeopteryx* is a partial, disarticulated skeleton, with wing feathers arranged in fan-like fashion about its arms and symmetrically arranged tail feathers emerging from each side of a long bony tail. This unique anatomy, intermediate between that of modern birds and fossil reptiles, caused experts to disagree about the importance of the fossil.

The story begins in November 1861, when German paleontologist Johann Andreas Wagner announced the discovery of a feathered reptile skeleton discovered in the Solnhofen Limestone of Langenaltheimer Haardt, Bavaria. Wagner was unable to see the fossil himself and relied on information provided by a lawyer, O. J. Witte, who had examined the fossil in the collection of local doctor Carl Häberlein. Only a few months earlier, Hermann von Meyer had announced

Name	*Archaeopteryx*
Discovered	Langenaltheimer Haardt, Bavaria, Germany, by anonymous quarryman and then passed to Dr. Carl Häberlein, 1861
Described	By Hermann von Meyer, 1861
Importance	The first Mesozoic feathered dinosaur to be discovered, and the oldest known bird
Classification	Saurischia, Theropoda, Coelurosauria, Maniraptora, Avialae

the discovery of a single fossil feather from the Solnhofen Limestone, essentially identical to those of modern birds, and had published the name *Archaeopteryx*. Whether von Meyer intended to use this name for the feather or for the skeleton owned by Häberlein remains a mystery. Partly as a snub to evolutionary scientists, Wagner gave Häberlein's fossil the rather nondescript name *Griphosaurus*, meaning "enigma lizard." He was opposed to the concept of evolution and slipped into his announcement the assertion that, although Darwin and his colleagues may well seize on *Griphosaurus* as evidence for evolution, they would be wrong to do so.

Describing the fossil

In 1862, Häberlein sold the fossil to the British Museum. It was thus in the hands of Richard Owen, and he immediately set about describing it. Owen noted that the robust wishbone suggested well-developed flight abilities, the wing proportions resembled those of falcons, and the foot was adapted for perching; neither of the latter two claims are seen as true today.

Throughout his description, Owen emphasized the similarity between *Archaeopteryx* and modern birds, downplaying the clawed, unfused fingers and long, bony tail clearly present in the specimen. Concluding that it was very much a modern bird, albeit one that was worthy of its own distinct taxonomic order, nowhere in his description did Owen refer to the idea that *Archaeopteryx* may have been an evolutionary intermediate between birds and reptiles.

The great evolution debate

Given that Darwin's *On the Origin of Species* was published in 1859, the timing of this discovery was uncanny. Some immediately recognized the importance of *Archaeopteryx* for evolutionary theory. In a letter to Darwin in 1863, Hugh Falconer wrote: "Had the Solnhofen quarries been commissioned to turn out a strange being à la Darwin, it could not have executed the behest more handsomely than with the *Archaeopteryx*."

Thomas Huxley—nicknamed "Darwin's bulldog"—criticized Owen's identifications of the fossil in 1868, arguing that the "right leg" was really the left, the "ventral surfaces" of the vertebrae were really the dorsal surfaces, and so on. Owen had mistakenly concluded that the fossil lay with its ventral surface facing the viewer—in fact, the opposite was true. However, although he later argued for a link between birds and dinosaurs, Huxley's views on *Archaeopteryx* were surprisingly uncontroversial: like Owen, he regarded it as a true bird.

The Berlin specimen

A second, more spectacularly preserved, *Archaeopteryx* specimen was discovered in 1876, and after it was obtained by Berlin's Humboldt Museum it became known as the "Berlin specimen." Fully articulated and with a complete skull, the Berlin *Archaeopteryx* soon overtook the London specimen in terms of iconic status and is today the most familiar of all the *Archaeopteryx* specimens.

Left The modern view of *Archaeopteryx* is of a small, dinosaur-like bird that was probably capable of running, climbing, and flying.

Above The fine-grained layers of the Solnhofen limestone preserve numerous fossils in exquisite detail. Invertebrates and fish are the commonest fossils.

35

Neglected
Scelidosaurus

By the 1840s, scientists generally recognized dinosaurs as distinct, but there was far less consensus about their appearance in the years to come. Owen saw them as heavy-bodied quadrupeds, whereas Mantell and Leidy argued that some dinosaurs were at least partially bipedal. Then, in the late 1850s, the discovery of a near-complete skeleton offered new information on posture and body shape.

James Harrison of Charmouth, Dorset, England, sent Owen a number of Lower Jurassic reptile bones he had found while quarrying the cliffs of Black Ven, near Charmouth in 1858. Owen recognized that the fossils belonged to a terrestrial dinosaur with some similarities to *Iguanodon*, and in 1859 he named the new animal *Scelidosaurus*, later adding the species name *harrisonii*. One of the bones, an incomplete femur, shows thick bone walls and large muscle attachment sites. These features so impressed Owen that he regarded its owner to be capable of "more vigorous action of the hind limbs" than other dinosaurs—hence the generic name, which means "limb lizard."

A properly illustrated description of the animal's skull did not appear until 1861. The rest of this specimen—an almost complete articulated skeleton, approximately 13 ft (4 m) long and encased in hard limestone blocks—was also uncovered, and described by Owen in 1863. He also described an assortment of other bones, including a knee joint and additional remains that he interpreted as having come from a juvenile or even foetal scelidosaur.

Bony defenses

The skull and virtually complete skeleton were encased in matrix when Owen studied them. However, he recognized that the jaws and teeth probably indicated a herbivorous lifestyle. He also described the rows of bony scutes—thickened horny plates—that run along the top of the neck and tail, and along the animal's sides, writing that the creature is "defended by several longitudinal series of massive dermal bones." These

Name	*Scelidosaurus harrisonii*
Discovered	Charmouth, Dorset, England, by James Harrison, 1858
Described	By Richard Owen, 1861
Importance	The first complete dinosaur skeleton to be discovered, and the first Jurassic ornithischian to be discovered
Classification	Ornithischia, Thyreophora

scutes were still articulated to the upper and lower surfaces of the tail vertebrae, and the well-preserved hind limbs and four-toed feet were also still articulated.

This impressive specimen remained preserved in its nodule until the 1960s, during which time staff of the then British Museum (Natural History) began the long process of removing the bones from the encasing sediment by way of acid preparation. The museum's fossil reptile expert, Alan Charig, planned to provide a thorough modern description of the specimen when this preparatory work was completed.

Owen's missed opportunity

Scelidosaurus is the first dinosaur for which a good, articulated skeleton came to light, and it is particularly interesting when seen in its historical perspective. Whereas Gideon Mantell imagined dinosaurs to be long-bodied and lizard-like, and Joseph Leidy proposed that they were kangaroo-like bipeds, Owen described them as heavy-limbed quadrupeds. Here, in *Scelidosaurus*, he had a fossil that could, potentially, have been used to demonstrate the at least partial correctness of his theory. However, he failed to make the most of this opportunity and why he did so remains a fascinating question.

David Norman has argued that, by the late 1850s and early 60s, Owen had become unable to keep abreast of everything: he was involved in a huge number of diverse projects; he was engaged in high-profile politicking in order to get his planned Museum of Natural History approved and built; and he had a crowded lecture schedule. He was also embroiled in several famous disputes with various other scientists at this time, including the "hippocampus debate" with Thomas Huxley. Moreover, the 1859 publication of *On the*

Origin of Species must have given Owen plenty to think about and respond to—he published an anonymous hostile review of the book in 1860, for example.

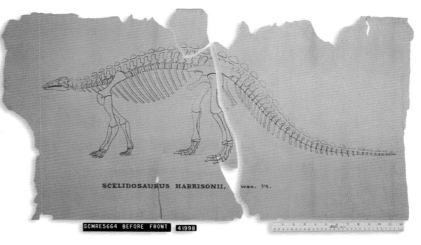

SCELIDOSAURUS HARRISONII, won. ¼.

SCMRE5664 BEFORE FRONT 41998

Scelidosaurus remained the earliest known recognized ornithischian until the poorly known heterodontosaurid *Geranosaurus* was named in 1911, and even after that date it long remained the only early species for which good remains were known. As a result, *Scelidosaurus* became widely used as the best example of a primitive ornithischian.

Type specimens

A type specimen is the individual specimen to which a species' name should be technically associated. During the 1880s, it was decided that the knee joint from Harrison's specimen should be regarded as the type specimen of *Scelidosaurus*. Unfortunately, this bone eventually proved to come from a theropod, as did the original femur that had inspired the name *Scelidosaurus*. This problem was finally resolved in 1992, when the name *Scelidosaurus* was officially transferred to the near-complete skull and skeleton.

Above The original scelidosaur fossil, shown here, was discovered in an articulated state and with most of its armor still in place.

Above This reconstruction of *Scelidosaurus* was produced by American paleontologist Othniel Marsh in 1895. **Below** During the 1850s and 60s, Richard Owen was kept busy with numerous projects on modern and fossil animals.

37

A new tiny dinosaur

Discovered in the Solnhofen Limestone by German physician and fossil collector Joseph Oberndorfer, the exquisite *Compsognathus* fossil is still regarded as one of Europe's most complete and best preserved theropods.

Above Thanks to new discoveries made in China, we now know that *Compsognathus* and its relatives were covered in hair-like "proto-feathers."

By 1860, the several fossil animals that were identified as part of Owen's Dinosauria had one thing in common: they were large. The smallest, *Scelidosaurus*, was about 13 ft (4 m) long. In 1861, the Solnhofen Limestone's spectacular small feathered *Archaeopteryx* specimen (*see* pp.34–35) was joined by a second exciting new creature, also less than 3 ft (1 m) long. Described by German paleontologist Johann Andreas Wagner, this fossil is one of Europe's most exciting discoveries.

Distorted in death

Wagner began his description by complaining that the skeleton was damaged in many places and preserved in a distorted posture such that the neck was twisted backward and the limbs overlapped one another—a common dinosaur death posture. Noting the long, delicate skull and very long, slender lower legs and feet, Wagner named the fossil *Compsognathus longipes*, meaning "dainty jaw, long feet."

Wagner was very aware of the marked difference in length between the hind and forelimbs, noting that the former are only half as long as the latter. Although this and other features demonstrate a bipedal posture for this animal, he did not discuss this further, so we cannot be sure how he imagined *Compsognathus* to look in life. However, a life-sized lithographic plate illustrating the specimen accompanied his 1861 description—and despite his initial complaints, he must still have regarded this as an important find.

Early interpretations

It is easy today to see in *Compsognathus* all the characteristic features of theropod dinosaurs: raptorial hands, bipedal posture, long hind limbs, and three-toed (in terms of function) feet. Although the theropods *Megalosaurus* from England and *Poekilopleuron* from France were described by this time, their anatomy was not well known enough for anyone to realize that they were giant relatives of *Compsognathus*.

Wagner regarded *Compsognathus* as a lizard, dismissing its superficially bird-like shape as unimportant. He did concede that certain features, such as the shape of the shoulder bones, resemble those of crocodilians and birds more than those of lizards; he also noted that the long pelvic bones differ from those of lizards and crocodilians, and most recall those of birds.

Evolutionary intermediate

Others were soon to argue that the strong similarity present between *Compsognathus* and birds was no coincidence. As early as 1864, the German anatomist Carl Gegenbaur pointed out that the animal appeared to be intermediate between reptiles and modern birds in the anatomy of its ankle bones. This made him the first person to hint at an evolutionary relationship between dinosaurs and birds.

In 1866, Edward Cope recognized *Compsognathus* as a dinosaur and created the new group Symphopoda for it—he did this because the tibia and fibula bones were united with the astragalus and calcaneum (heel bone) in the ankle, and this was a feature hitherto thought to be unique to birds. Drawing attention to the long neck vertebrae, lightly constructed skull, and ankle anatomy, he regarded the species as somehow close to birds, and particularly to the bird groups that he regarded as the most primitive—the penguins and ratites. Later, the *Compsognathus* ankle bone configuration that fascinated Cope was shown to be widespread throughout dinosaurs.

Writing in 1869 about the affinity he perceived between dinosaurs and birds, Thomas Huxley regarded *Compsognathus* as part of the Dinosauria, or at least as close to this group. He later gave it its own near-dinosaurian group, Compsognatha. Thanks to the then brand-new English dinosaur *Hypsilophodon*, Huxley knew that at least some dinosaurs have long pubic bones that project posteriorly, very much like those of birds; as a result he imagined that *Hypsilophodon*

and its relatives were evolutionary intermediates between more primitive dinosaurs and the birds. Although we now know that *Compsognathus* had a forward-projecting pubis with a large "foot" at its end, the exact conformation of the pubis was not clear to Wagner and Huxley. The latter must have thought that the pubis in *Compsognathus* was not posteriorly directed like that of *Hypsilophodon*, however, because he noted that *Hypsilophodon*-like pelvic bones are "by no means universal" among dinosaurs.

By the 1890s, *Compsognathus* was properly associated with other predatory dinosaurs and included in the Theropoda, a group first named by Marsh in 1881. However, it remained one of the best-known small theropods throughout the rest of the 19th century and for much of the 20th. As such, it was destined to play a major role in discussions about the origins of birds.

Above The spectacular Solnhofen skeleton of *Compsognathus* shows the long hind limbs and slender, bird-like neck and skull now known to be typical for many predatory dinosaurs.

Hypsilophodon and the bird-dinosaurs

The ornithopod now known as *Hypsilophodon foxii* was discovered in 1849 when men working at Cowleaze Chine on the Isle of Wight, off the south coast of England, discovered an articulated partial skeleton.

Above It was immediately apparent that *Hypsilophodon* was a herbivore, because it had a narrow, pointed beak and leaf-shaped cheek teeth adapted to eating vegetation.

Broken into two blocks, it went to two different collections: those of Gideon Mantell and James Scott Bowerbank, and in 1849 Mantell described its neck vertebrae as a young *Iguanodon*. In 1855, Richard Owen published his description of both parts of the specimen, and, like Mantell, he maintained that it was a juvenile *Iguanodon*.

A second specimen

In January 1868, the Reverend William Fox discovered another specimen, this time consisting of a skull and some vertebrae. Thomas Huxley was able to examine it and, with Fox's permission, he published a brief note on it in 1869 and then a longer paper in 1870. Naming it *Hypsilophodon foxii*, he showed that its vertebrae were nearly identical to the vertebrae of the Mantell-Bowerbank specimen. Accordingly, the 1849 find "could not possibly be *Iguanodon*," and must also belong to *H. foxii*. Owen disagreed with Huxley's newly proposed distinction and continued to regard *H. foxii* as nothing more than a juvenile *Iguanodon*. He pointed to the form of the teeth and lower jaw as evidence.

Hypsilophodon translates as "high-crested tooth," so it has often been assumed that this was Huxley's intended meaning of the name. In fact, he was naming the new dinosaur

Right Thanks to several articulated *Hypsilophodon* skeletons discovered over the years, scientists have been able to produce this three-dimensional reconstruction.

40

after *Hypsilophus*, a modern iguana. Given that *Hypsilophodon* was previously regarded as a small version of *Iguanodon* (which of course means "iguana tooth"), it seems likely that Huxley was maintaining an iguana-based "theme" in coining this new name.

Hypsilophodon was central to Huxley's case that dinosaurs included the ancestors of birds. He had already been a champion of a dinosaurian origin for birds, but new data from *H. foxii* helped him to strengthen his case: he was able to show that two bones on the Mantell-Bowerbank specimen, interpreted by Owen as a tibia and fibula, were in fact the pubis and ischium. These bones pointed backward in *Hypsilophodon*, just as they do in birds. Huxley was so impressed with this fact that he even said that had the pubis and ischium of *Hypsilophodon* been discovered on their own, "they would have been unhesitatingly referred to Aves."

Hypsilophodon reconstructed

Additional *Hypsilophodon* specimens were described by John W. Hulke during the 1870s and 80s. In 1882, Hulke provided the first skeletal reconstruction of the dinosaur. Although this can be considered superficially accurate, he assumed a flexible tail, short neck, and hind limbs kept strongly bent, with the feet mostly flat on the ground—all details that have since proved inaccurate. He imagined *Hypsilophodon* to be quadrupedal and, superficially, not dissimilar from a large living lizard.

It was this vision of *Hypsilophodon* that became best known to the public of the time, in part thanks to a life restoration that appeared

in one of the most influential early books on prehistoric animals, Henry Neville Hutchinson's *Extinct Monsters* of 1894. The illustration in that book, produced by Dutch artist Joseph Smit, showed *Hypsilophodon* walking quadrupedally but also able to sit up in an erect posture. Given Huxley's earlier emphasis on the bird-like attributes of this dinosaur, it is ironic that these new reconstructions made it more lizard-like.

The presence of a four-toed foot in *Hypsilophodon* inspired Hulke to hint at a relationship with *Scelidosaurus* (*Iguanodon* was known by then to be three-toed), but the pointed, curved claws were taken as evidence that "*Hypsilophodon* was adapted to climbing upon rocks and trees." The idea that the dinosaur was at least partially quadrupedal, and an able climber, persisted until the late 1960s.

Changing views

Views on the appearance of *Hypsilophodon* first began to change in 1895, when Othniel Marsh published a new skeletal reconstruction. He could do this thanks to American fossils of the *Hypsilophodon*-like dinosaur *Laosaurus* discovered in the Morrison Formation. Marsh depicted *Hypsilophodon* as an erect-limbed biped. By this time it was clear that predatory dinosaurs were similar in posture, and although some dinosaurs had proved to be quadrupedal, the pachydermal giants imagined by Richard Owen disappeared altogether from the scientific literature at this time.

Above Its long, slender feet and body proportions show that *Hypsilophodon* was an agile biped adapted to life on the ground. It was certainly not a climber, as was once thought.

Digging up the dead

Finding a fossil is the start of what is often a long, grueling process. Some fossils can be lifted from the ground straight away, but it is more typical for excavation to be difficult. When fossils project from the side of a cliff or riverbank, the layers of sediment on top—known as the overburden—must be removed first, sometimes requiring the use of heavy machinery, and tools such as jackhammers. The sediments that entomb fossils vary from soft to hard and resistant—some excavation teams have even used dynamite to remove overburden.

The proper preparation of a fossil can be done only in a laboratory or workshop. They are typically discovered in remote places, and transporting them back to the laboratory can be difficult and expensive.

Right At Dinosaur National Monument on the Colorado/Utah border, the bones of sauropods and other dinosaurs have been carefully prepared from the rock face, mostly using hammers and other such tools.

Below left A diverse tool kit is required for fossil excavation. Hammers, picks, and spades are needed as well as delicate implements such as brushes, chisels, and dentistry tools.

Below right Large fossils, such as this *Tyrannosaurus rex* pelvis are incredibly heavy. It requires a whole team of people, or even lifting cranes, to move such objects.

American stegosaurs: the "roofed reptiles"

Stegosaurs, the plated dinosaurs, look unmistakable. Their pillar-like hind limbs are much longer than their forelimbs, their hips are tremendously broad, and their bodies are decorated with spines and, in some species, large triangular or diamond-shaped plates.

Below *Stegosaurus* was discovered in the rocks of the Morrison Formation. Here, Dr. Robert Bakker and other fossil hunters search for new Morrison dinosaur discoveries.

Our modern view of stegosaurs emerged after 1891, when Othniel Marsh produced the first skeletal reconstruction. Before Marsh's work there was a great deal of uncertainty about what these amazing dinosaurs looked like, how they lived, and what they were.

DISCOVERY PROFILE

Name	*Stegosaurus armatus*
Discovered	Near Morrison, Colorado, U.S.A., by Arthur Lakes, in 1877
Described	By Othniel Charles Marsh, 1877
Importance	The first North American stegosaur, and first stegosaur with armor plates preserved
Classification	Ornithischia, Thyreophora, Stegosauria, Stegosauridae

Naming *Stegosaurus*

Stegosaurus armatus, the first of many *Stegosaurus* species to be coined, was named by Marsh in 1877 for material collected just north of Morrison, Colorado. One of his salaried collectors, Arthur Lakes, excavated the bones earlier that year.

Consisting of vertebrae and partial limb and pelvic bones, the remains also included some of the distinctive bony spinal plates, including one that was more than 3 ft (1 m) in length, albeit broken into fragments. Numerous cylindrical teeth were also found and attributed to the specimen, but these later turned out to belong to the sauropod *Diplodocus*.

Marsh imagined the plates to be similar to the large, flat bones that form the shells of certain fossil sea turtles, and he proposed that they were supported by the tall neural spines present on the vertebrae. Presumably inspired by this perceived similarity to giant turtles, he wrote that *Stegosaurus* moved mainly by swimming and regarded it as having affinities with dinosaurs, plesiosaurs, and turtles. He described this bizarre new reptile as "one of the most remarkable animals yet discovered."

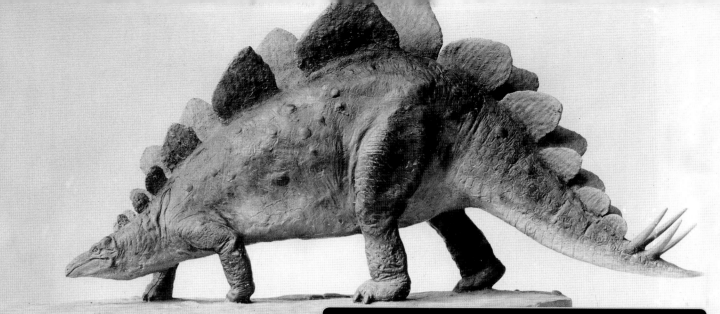

Marsh received a great deal more *Stegosaurus* material from the Morrison Formation, including a partial *Stegosaurus* skull. This material, and the fact that he already knew about the size disparity between the hind and forelimbs, made him confident by 1879 that *Stegosaurus* was a dinosaur. Marsh's great rival, Edward Cope, also received Morrison Formation stegosaur material about this time, and in 1878 named the new species *Hypsirophus discurus*. Although Cope's *Hypsirophus* material also included some theropod bones, it is now agreed that *H. discurus* is synonymous with *S. armatus*. Marsh argued this in 1882, but Cope ignored him and continued using the name *Hypsirophus* until his death.

Anatomical puzzles

Marsh gave up the idea that these dinosaurs might be turtle-like. However, he still regarded them as predominantly aquatic, arguing that the solid limb bones lent support to this idea (aquatic vertebrates, such as plesiosaurs and sea cows, often have very thick-walled or solid bones).

By 1880, Marsh regarded the different lengths of the hind and forelimbs as proof that *Stegosaurus* was "more or less bipedal in its movements on land." Also of interest was his comment in 1881 that the forelimbs are so powerfully muscled that they must have served some purpose other than locomotion, and he suggested that *Stegosaurus* was perhaps able to use its arms in foraging or when defending itself.

Marsh could not decide how the large plates were arranged in the living animal, and similar uncertainty existed about the large spikes that had also been found with these animals. Perhaps the plates were arranged in several symmetrical rows on either side of the back and tail, and the spikes protected the forelimbs, he suggested.

Marsh's first reconstruction

The true appearance of *Stegosaurus* was revealed in 1885, when a specimen that became known as the "road kill" *Stegosaurus* was discovered near Cañon City, Colorado, by M. P. Felch. Near-complete, lying on its side, and squashed flat, it showed that the plates formed a row along the backbone, arranged in an overlapping pattern. This specimen also revealed that numerous small ossicles protected the throat region. Initially named as a new species, *S. stenops*, it is now held synonymous with *S. armatus*.

By 1891, Marsh knew enough about *Stegosaurus* to attempt a complete skeletal reconstruction. He gave the animal an arching back, a neck that sloped down, a tail that drooped down to the ground, and plates arranged in a single row along the top of the backbone. While known to be inaccurate in tail posture and other details, this reconstruction was still being used until very recently.

Above This model, produced by Charles Knight in 1903, shows the humped back and low head position once thought accurate for *Stegosaurus*.

The Cretaceous
bison

Above *Triceratops* was one of the largest of the horned dinosaurs, exceeding 23 ft (7 m) in length and with a skull up to almost 8 ft (2.4 m) long.

The gigantic Late Cretaceous horned ceratopsian *Triceratops*, one of the most iconic of all dinosaurs, is today known from spectacular skulls and near-complete skeletons. The initial remains of this dinosaur, however, were very fragmentary and identified incorrectly.

Above Othniel Marsh (top, center) appears here in 1872 surrounded by his armed field crew. Marsh's strong rivalry with Edward Cope put their field crews into competition with each other, although stories that they sometimes came to blows are probably embellishment.

The story of *Triceratops* begins during the 1880s, when George Cannon and several of his colleagues working for the U.S. Geological Survey discovered fragmentary bones and teeth in the Denver Basin. In 1887, Cannon excavated a partial skull roof to which were attached a pair of slender horns 2 ft (60 cm) long.

The horns were sent to Othniel Marsh, the Survey's official vertebrate paleontologist, who thought they were reminiscent of those in cattle. As a result Marsh named the remains *Bison alticornis*. In late 1887, he went on to describe the species as "one of the largest of American bovines, and one differing widely from those already described." Some fossil bison have far

longer, far straighter horns than we are used to in modern bison, so Marsh's identification is not unreasonable. This was, after all, the first clear indication of a gigantic horned Cretaceous reptile, and Marsh was later to note how the hollow bases and blood vessel patterns of the *B. alticornis* horns strongly resembled the same features of bison horns.

The mistake Marsh made was understandable when one considers that the exact age of the rocks that yielded these fossils was controversial. During the late 1880s, stratigraphic information on the western United States had only just started to accrue. Vertebrate paleontologists regarded the rocks of the Denver region as Cretaceous in age, while paleobotanists had argued that they were post-Cretaceous. Marsh's "fossil bison" confused things further, since it caused him to suggest a late Pliocene age for the rocks.

A family is born

Additional horns were discovered by 1888, this time from Montana. Marsh recognized that they belonged to a massive, *Stegosaurus*-like reptile. He named the fossils *Ceratops montanus* and created a new family for them: Ceratopsidae. Marsh thought that not only were ceratopsids similar in appearance to stegosaurs, they were also close relatives, and he even regarded *Stegosaurus* as the ancestor of the Cretaceous horned dinosaurs.

At about the same time, Marsh's collector John B. Hatcher was confronted with a partial horned dinosaur skull in Wyoming, and by 1889, Marsh had been sent the first of many specimens. He initially named the new Wyoming fossil *Ceratops horridus*, but by August 1889, Marsh decided that it was worthy of its own genus, and the name *Triceratops*, or "three-horned face" was born. Besides drawing obvious attention to the three-horned anatomy of the *Triceratops* skull, Marsh described how a projecting bone at the front of the upper jaw (he termed this the rostral bone) formed the top half of a huge horny beak. The presence of a massive frill surrounding the neck was also evident and was inferred to provide "strong protection to the neck."

It was by this time clear to Marsh that *Bison alticornis* was in fact a ceratopsid similar to *Triceratops*, and was thus Cretaceous in age, as were *Ceratops montanus* and *Triceratops*. This ended his earlier theory that the Denver rocks might be late Pliocene in age. Instead of including *Bison alticornis* within *Triceratops*, Marsh came to regard it as part of *Ceratops*. However, because these remains lack any features that allow them to be identified to genus level, the name *Ceratops alticornis* is generally ignored today.

How many species?

In 1889, Marsh added two further species to the genus *Triceratops*: *T. flabellatus*, from Wyoming; and *T. galeus*, from Colorado, and by the early decades of the 20th century, a total of 16 *Triceratops* species were known. Nearly all of these proposed species came from the same geographical area and were very similar in geological age. The differences between them are minor and relate to the size and shape of the horns and frill. Experts today recognize only two species: *T. horridus* and *T. prorsus*, which has a shorter snout and horns.

Above With its solid neck frill and three horns, *Triceratops* is one of the most distinctive of all dinosaurs.

DISCOVERY PROFILE

Name	*Triceratops horridus*
Discovered	Wyoming, U.S.A., by John B. Hatcher, *c.*1888
Described	By Othniel Marsh, 1889
Importance	One of the first indications of gigantic horned dinosaurs
Classification	Ornithischia, Marginocephalia, Ceratopsia, Ceratopsidae

47

North America and the age of sauropods

Sauropods (giant, long-necked dinosaurs) were first recognized in the 1840s due to the discovery of *Cetiosaurus*, but no one had a clear idea of how they looked. Then, in the 1870s, spectacular discoveries in North America demonstrated the real appearance of these amazing giants.

Above The skull of *Camarasaurus* was first revealed to science during the 1880s. It is short and deep, with enormous nostril openings and stout jaws.

North American sauropod material was described for the first time in 1877, when a flurry of new announcements revealed the existence of several new kinds of sauropod in the Upper Jurassic Morrison Formation. Othniel Marsh described *Atlantosaurus* and *Apatosaurus*, and Edward Cope named *Camarasaurus*, *Amphicoelias*, and *Dystrophaeus*. Due to the intense rivalry that existed between Marsh and Cope, they rushed their new dinosaurs into print, and their initial announcements were extremely terse and often lacking illustrations. Some of the dinosaurs they named were based on poor material and sank into obscurity. However, among the best of their new sauropods was Cope's *Camarasaurus supremus*, discovered by Oramel Lucas near Cañon City, Colorado, in the same quarry as the "road kill" *Stegosaurus* specimen (*see* pp.44–45).

Camarasaurus preserved vertebrae, limb girdles, and limb bones, making it the best known sauropod. In 1877, John Ryder used the fossils to produce the first sauropod reconstruction. His illustration correctly showed *Camarasaurus* as long-necked and quadrupedal but included a hypothetical skull and made the tail too deep and paddle-like. It was also a composite, incorporating the remains of as many as six different individuals. Exhibited at a meeting of the American Philosophical Society, Ryder's reconstruction was produced at life-size, over 50 ft (15 m) in length.

Interpreting sauropod anatomy

The large amount of *Camarasaurus* material that Cope received allowed him to characterize most details of the sauropod skeleton. He commented on the complex, strut-like laminae and large, bony concavities on the sides of the huge, lightly constructed, hollow vertebrae, noting that the vertebrae were "lighter in proportion to their bulk than in any air-breathing vertebrate." The pillar-like limb bones of *Camarasaurus* were surprisingly long, and Cope was inspired by these and the long neck to suggest that *Camarasaurus* was giraffe-like and fed from the tops of trees.

Interestingly, in 1877 Marsh and Cope believed their new animals were terrestrial, whereas by the following year—when Marsh published the name Sauropoda—they both regarded sauropods to be

---

DISCOVERY PROFILE

Name *Camarasaurus supremus*

Discovered Near Cañon City, Colorado, by Oramel W. Lucas, 1877

Described By Edward Cope, 1877

Importance The first North American sauropod and first sauropod used to create a skeletal reconstruction

Classification Saurischia, Sauropodomorpha, Sauropoda, Macronaria

---

Right Excellent fossils have made *Camarasaurus* the best known sauropod. It was medium-sized for a sauropod, at about 50 ft (15 m) in length, and was equipped with a stout neck and medium-length tail.

Above For years, this *Apatosaurus* skeleton in New York was mounted with an incorrect, camarasaur-like skull, leading people to think that *Apatosaurus* had a blunt, short head.

Right The rocks of the Morrison Formation by Colorado's Purgatoire River preserve abundant sauropod trackways.

water dwellers. Why these authors changed their mind is perplexing. In a 1975 article on this issue, Walter Coombs proposed that the reason might have been that both men read Owen's articles on *Cetiosaurus* (*see* pp.30–31) and were ultimately led astray by his interpretation. It is difficult to be sure that this is really what happened, because Marsh and Cope cited geological and anatomical evidence in support of their aquatic interpretation of sauropods.

A golden age

The late 1870s were undoubtedly a rich time for dinosaur discovery in North America, and the fauna of the Upper Jurassic Morrison Formation proved diverse and exciting. The sauropods *Camarasaurus*, *Apatosaurus*, and *Diplodocus* were all described within a few years of each other, as were *Stegosaurus* and the large theropod *Allosaurus*.

In 1878, Marsh named *Diplodocus* for another find at the Cañon City quarry, and *Brontosaurus* was named in 1879 for an excellent skeleton found in Wyoming. However, the species concerned, *B. excelsus*, was eventually classified in the same genus as the type species of *Apatosaurus*, *A. ajax*. The new *Brontosaurus* material allowed Marsh to produce the second skeletal reconstruction of a sauropod. Far more accurate than Ryder's version, it showed a better proportioned tail and neck and also used a real sauropod skull rather than a hypothetical one. However, the skull was actually that of a *Camarasaurus* found some miles away.

THE MORRISON FORMATION

Cope and Marsh's sauropods were from the Morrison Formation, an Upper Jurassic rock unit that extends over 600,000 square miles (1.5 million sq km) of the western United States. Between the 1870s and early 1900s, an extraordinary variety of Morrison dinosaurs— giant sauropods, theropods, stegosaurs, and others—were discovered and sent to the museums in the east. These had a huge influence on the scientific understanding of dinosaurs.

Below The Morrison Formation is one of the most fertile dinosaur-bearing rock units in the world.

49

The Great Dinosaur Rush

The beginning of the 20th century saw the emergence of a global golden age in dinosaur discovery. Driven by a quest for spectacular museum pieces, teams of fossil hunters in Canada and the United States discovered amazing new dinosaurs. The gigantic predator *Tyrannosaurus* was discovered at this time, as were a dazzling array of horned and crested herbivores. Exploration in Africa and Asia also led to major discoveries. The public were amazed and awed by these new museum pieces, and illustrations and photos of these specimens became iconic.

Facing page Barnum Brown, one of the key characters in the dinosaur rush of the early 1900s, is shown here excavating dinosaur bones at Big Horn, Wyoming, in 1934. Brown discovered numerous Cretaceous dinosaurs in the American West, including *Tyrannosaurus*, *Ankylosaurus*, and many others.

CANADA

Little Sandhill Creek

Red Deer River

Tolman Ferry | Morin
Berry Creek
Sandhill Creek

Dead Lodge Canyon

CANADA

Red Deer River

Montana
Wyoming
Colorado
Oklahoma

New Mexico

SOUTH
AMERICA

Neuquén Province

Río Negro Province
Chubut Province

The early 20th century

Above During the early decades of the 20th century, many entirely new kinds of dinosaurs—including *Corythosaurus*, shown here—were discovered in the Late Cretaceous rocks of western North America.

The spectacular North American dinosaur discoveries of the 1870s, 80s, and 90s made the continent's western interior a major focus for finding new specimens, and this golden age in North American dinosaur discovery was still well underway in the early years of the 20th century. However, other places around the world soon began to reveal their dinosaur fossils as well.

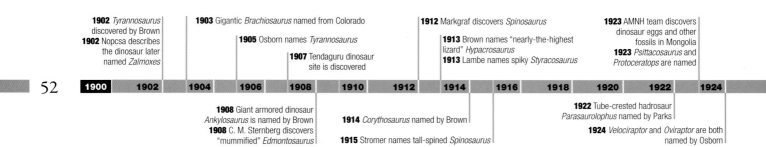

1902 *Tyrannosaurus* discovered by Brown
1902 Nopcsa describes the dinosaur later named *Zalmoxes*

1903 Gigantic *Brachiosaurus* named from Colorado
1905 Osborn names *Tyrannosaurus*
1907 Tendaguru dinosaur site is discovered

1912 Markgraf discovers *Spinosaurus*
1913 Brown names "nearly-the-highest lizard" *Hypacrosaurus*
1913 Lambe names spiky *Styracosaurus*

1923 AMNH team discovers dinosaur eggs and other fossils in Mongolia
1923 *Psittacosaurus* and *Protoceratops* are named

52 | **1900** | 1902 | 1904 | 1906 | 1908 | 1910 | 1912 | 1914 | 1916 | 1918 | 1920 | 1922 | 1924

1908 Giant armored dinosaur *Ankylosaurus* is named by Brown
1908 C. M. Sternberg discovers "mummified" *Edmontosaurus*

1914 *Corythosaurus* named by Brown
1915 Stromer names tall-spined *Spinosaurus*

1922 Tube-crested hadrosaur *Parasaurolophus* named by Parks
1924 *Velociraptor* and *Oviraptor* are both named by Osborn

UNITED
KINGDOM
Isle of Wight
Trossingen
Ha eg Basin
El-Bahariya
AFRICA
Tendaguru
Mwakasyunguti, Malawi
Mount Fletcher, Cape Province
Mount Fletcher, Cape Province

MONGOLIA
Shabarakh Usu, Gobi Desert
Tsagan Nor Basin, Gobi Desert
CHINA
INDIA
Jabalpur, Madhya Pradesh
Heilongjiang (Laiyang), Shandong
Ningxiachou, Shandong

Clutha, Queensland
AUSTRALIA
Durham Downs, Queensland
Lightning Ridge, New South Wales

The 20th century opened with the naming of the creature that soon became the most famous of all dinosaurs: *Tyrannosaurus rex*, from the Late Cretaceous of western North America. Its discovery was a continuation of fieldwork by Barnum Brown, who in 1897 was hired by Henry Osborn of the American Museum of Natural History. Brown collected fossils in the Jurassic rocks of Wyoming and the Upper Cretaceous rocks of Montana, and, from 1910 to 1915, he prospected the Red Deer River region in Alberta, discovering several new Canadian dinosaurs. Canadian dinosaurs were also excavated during the same period by Charles

H. Sternberg and his three sons George, Charles M., and Levi. Both Brown and the Sternbergs used rafts and motorboats while traveling along the Red Deer River in search of fossils.

The African connection

Stunning discoveries were made in Africa, too. In 1907, a location in Deutsch Ostafrika (now Tanzania) revealed a phenomenal wealth of dinosaur remains. This site, known as Tendaguru, was soon exploited by successive expeditions from Berlin's Museum für Naturkunde. In 1914, German paleontologist and geologist Werner Janensch named the Tendaguru sauropods *Dicraeosaurus* and *Brachiosaurus brancai*.

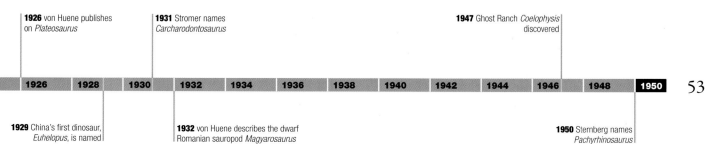

1926 von Huene publishes on *Plateosaurus*

1931 Stromer names *Carcharodontosaurus*

1947 Ghost Ranch *Coelophysis* discovered

| 1926 | 1928 | 1930 | 1932 | 1934 | 1936 | 1938 | 1940 | 1942 | 1944 | 1946 | 1948 | **1950** |

53

1929 China's first dinosaur, *Euhelopus*, is named

1932 von Huene describes the dwarf Romanian sauropod *Magyarosaurus*

1950 Sternberg names *Pachyrhinosaurus*

Facing page, top In this scene from the early 1900s, Barnum Brown and his excavation team prepare to move the plastered bones of *Tyrannosaurus* from Hell Creek, Montana.

Facing page, bottom In 1923, an American Natural History Museum team discovered these dinosaur eggs—identified at the time as belonging to *Protoceratops*—in the Gobi Desert.

Stegosaurs, ornithopods, and theropods were also discovered there. These finds showed that Africa's Late Jurassic fauna was broadly similar to that of contemporaneous North America.

Elsewhere in Africa, collector Richard Markgraf and paleontologist Ernst Stromer von Reichenbach recovered Late Cretaceous dinosaurs from El-Bahariya in the Great Western Desert of Egypt. The new specimens were described by Stromer between 1914 and 1936, including the giant *Spinosaurus* and *Carcharodontosaurus*. As suggested by its name, *Spinosaurus* had remarkably tall neural spines on the vertebrae of its back, though it was otherwise thought to be similar in appearance to other large theropods. Although the dinosaurs of Tendaguru and El-Bahariya stole the limelight, African dinosaurs were also reported from Malawi (then Nyasaland) and South Africa during the 1920s.

The Asian frontier

At about this time, another continent began to reveal its dinosaurs to the scientific world: Asia. One of the most important research efforts was a series of expeditions into the Gobi Desert by teams from the American Museum of Natural History (ironically, the scientists were looking for fossil mammals, not dinosaurs). Osborn thought that Mongolia might be the place where modern humans evolved, and Roy Chapman Andrews—a self-styled explorer more interested in modern mammals—was the team's leader. In 1923, they discovered the small theropods *Velociraptor* and *Oviraptor* and the early ceratopsians *Psittacosaurus* and *Protoceratops*. *Protoceratops* was clearly close to the ancestry of the great ceratopsids of North America, and the small theropods represented new, bird-like groups of uncertain affinity. Famously, dinosaur eggs and nests were discovered, too. India began producing dinosaur fossils in the 1860s, but an important haul of new Late Cretaceous theropods was reported in 1933.

The dark age

Despite the amazing discoveries of the 1920s and earlier, research on dinosaurs began to tail off during the middle decades of the 20th century, and the field entered what has been called the "dark age" or "quiet phase." Vertebrate paleontology became focused on mammals such as horses and rodents, and by the 1950s research on dinosaurs had all but stopped. Following the dinosaur rush of the late 1800s and early 1900s, it seemed that North American's potential for new dinosaurs had been exhausted. The great museum halls of North America and Europe no longer required spectacular showpieces such as *Tyrannosaurus* or *Diplodocus*. Gerhard Heilmann's *The Origin of Birds*, published in 1926, showed to the satisfaction of many scientists that dinosaurs did not include the ancestors of birds, as Huxley had argued during the 1860s, and dinosaurs were generally regarded as evolutionary "dead-ends" unworthy of scientific attention.

Below This mummified *Edmontosaurus*, discovered in 1908 by George and Levi Sternberg in Wyoming, was one of several hadrosaur mummies discovered during the "dinosaur rush."

King tyrant lizard

Below Barnum Brown discovered more than one *Tyrannosaurus* specimen, and in 1908 he discovered this one. It remains one of the best examples of its species and is today on display in New York.

In 1905, Henry Fairfield Osborn described a creature that soon became the most famous dinosaur of them all. The new species—a theropod—was collected on an American Museum of Natural History expedition led by Barnum Brown. Osborn had sent Brown to Montana to find a *Triceratops* for display, but the giant predator discovered on the trip astounded Osborn and his colleagues.

IDEAS ABOUT POSTURE

There is a widely held misconception that paleontologists during the early 20th century invariably believed that theropods and other dinosaurs walked upright with erect bodies and dragging tails. However, artwork produced at the time suggests that scientists also imagined them moving (at least occasionally) with horizontal bodies and tails. For example, although the much reproduced skeletal reconstruction and mounted skeleton created under Osborn's supervision show *Tyrannosaurus* standing in a diagonal posture, in Erwin Christman's detailed drawings of the vertebral column the back and tail are nearly horizontal.

Osborn named the specimen *Tyrannosaurus rex*—"king of the tyrant lizards"—and estimated its length at 39 ft (12 m) and its standing height at 19 ft (6 m). He imagined *Tyrannosaurus* stood with its back held diagonally and its tail approaching the ground, so the latter dimension was therefore the "rearing height" of the animal. Although the skull was still being prepared when Osborn wrote his article, he observed low, rounded horns above and in front of the eyes. He also noted the discovery of a large, long upper arm bone found alongside the *Tyrannosaurus* bones. Brown was confident that this belonged to the tyrannosaur, but Osborn was skeptical, and since then it has indeed been found to be part of a herbivorous dinosaur.

Other giant bones

Evidence for giant theropods had already been discovered in the latest Cretaceous of the Rocky Mountain West, but the fossils concerned were fragmentary and of uncertain affinities. In 1896, Marsh named *Ornithomimus grandis* for various bones from Wyoming, and, in 1892, Cope named *Manospondylus gigas* for two gigantic dorsal vertebrae. It is now generally thought that these were *T. rex*

Left Reaching 40 ft (12 m) in length and perhaps weighing as much as 11 tons (10 tonnes), *Tyrannosaurus* was massive and heavily built. The function of its short arms and two-fingered hands remains contentious.

vertebrae. However, due to the poor quality of the material it is not appropriate to replace the name *Tyrannosaurus rex* with the older *D. grandis* or *M. gigas*.

A second specimen

Tyrannosaurus rex was not the only huge theropod described in Osborn's 1905 article: he also named *Dynamosaurus imperiosus* for a specimen collected in 1900 near the Cheyenne River, Wyoming. It differs from *Tyrannosaurus* in the irregularly shaped bony plates on its back and sides. On the basis of this supposed difference, *Tyrannosaurus* and *Dynamosaurus* were separated.

It is now clear that the plates belong to an ankylosaurid. In fact, because they are marked with small craters that resemble tyrannosaur tooth marks, it is even possible that they are part of the tyrannosaur's stomach contents. By 1906, Osborn realized that *Dynamosaurus* and *Tyrannosaurus* are the same animal, but continued to think that *Tyrannosaurus* had armor plates.

Linear theories of evolution

Tyrannosaurus is particularly interesting from a historical perspective, because Osborn used it to help support a view of evolution termed orthogenesis. This is the idea that there is a predetermined trend within living things to move toward a certain "goal," and that the evolution of a group is therefore a direct process centered on the accentuation of one key detail. In the case of the great carnivorous

dinosaurs, it was thought that megalosaurs, allosaurs, and tyrannosaurs were members of a single lineage, and that *Tyrannosaurus* was the ultimate "end point" of an evolutionary trend toward greater size. In a 1917 article, Osborn described *Tyrannosaurus* as: "The most superb carnivorous mechanism among the terrestrial Vertebrata, in which raptorial destructive power and speed are combined; it represents the climax in the evolution of a series."

Osborn also noted, however, that *Tyrannosaurus* is similar to the smaller, far more lightly built ostrich dinosaurs. His iconic reconstruction of *Tyrannosaurus* equipped it with three fingers, but the discovery of a two-fingered hand in *Gorgosaurus* (in which the third digit is reduced to a vestige and only the metacarpal is present) raised the possibility that *Tyrannosaurus* might prove to be two-fingered as well. This was confirmed in 1988, when a *T. rex* preserving complete arms was finally discovered.

Below Osborn imagined *Tyrannosaurus* stood with a diagonal backbone and a dragging tail. For decades, Brown's 1908 *Tyrannosaurus* specimen was displayed in this posture.

The giant fused lizard

On a 1906 American Museum of Natural History expedition to the Hell Creek Beds of Montana, Peter Kaisen discovered the skull and skeleton of a remarkable new armored dinosaur. Kaisen collected for Barnum Brown, and in 1908 Brown described the animal, giving it the name *Ankylosaurus magniventris*. The animal was huge, at up to 30 ft (9 m) in length, and tank-like.

Above *Ankylosaurus* armor scutes, such as this, are keeled and oval-shaped. The scutes were arranged in rows along the animal's back, sides, and tail.

Below *Ankylosaurus* was a spectacular animal. It reached 30 feet (9 m) in length and bristled with spikes, plates, and horns.

Brown thought *Ankylosaurus* was a new kind of stegosaur. The specimen was reasonably complete, and he was able to reconstruct the entire animal, albeit incorrectly in parts. Brown had collected *Ankylosaurus* material before, when excavating the *Tyrannosaurus* specimen originally named *Dynamosaurus imperiosus*. Indeed, in his description of *Ankylosaurus*, he compared its scutes with those recently ascribed to *Tyrannosaurus* by Osborn. He concluded that they were quite different: those of the theropod were far thicker, less regular in shape, and different in surface texture. Ultimately, Brown's conclusions proved incorrect, because the scutes found with *Tyrannosaurus* were actually from *Ankylosaurus*.

Impressive armor

Kaisen's new specimen included a skull as well as limb bones, the shoulder girdle, ribs, vertebrae, and armor plates. The skull, which looks broad and almost triangular when seen from above, is covered in small bony plates and has small horn-like projections that point sideways and backward from the rear corners. Brown also noted that the isolated *Troodon* tooth described by Joseph Leidy in 1856 bears a certain resemblance to the *Ankylosaurus* teeth—but this is coincidental and the similarity is not strong.

The massive, short, wide vertebrae and long ribs showed that *Ankylosaurus* was a heavily built, broad-bodied animal, and its thick-boned shoulder girdle indicated that it had robust limbs. Among the unusual skeletal features Brown noted is the fusing of the ribs at the rear end of the rib cage to the adjacent vertebrae. *Ankylosaurus* was clearly protected by rows of oval scutes that cover much of the animal's dorsal surface, and Brown

ELEPHANTINE FEATURES

Brown noted that the *Ankylosaurus* skull contains symmetrically arranged bony chambers, similar to the sinuses seen in modern elephants. In elephants, these air-filled chambers serve to lighten the massive head, but their function in *ankylosaurs* remains a mystery even today.

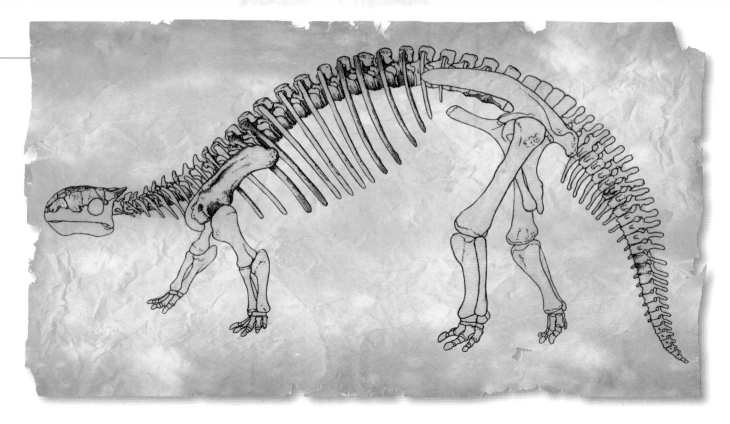

described how these differ in shape, presumably because different scutes belong to different parts of the body. Some are like an inverted V in cross-section, and some of the larger ones are almost flat. Two are united along one edge, and Brown noted that they resemble a similar structure identified in *Stereocephalus*—named from the older Judith River Beds by Henry Osborn and Lawrence Lambe in 1902—as part of a collar-like structure on the neck. He suggested that rings of armor plates might have encircled the tail, perhaps forming bony rings similar to those in some of the glyptodonts, a group of fossil, armadillo-like mammals.

A new family

Ankylosaurus seemed to be such a distinct specimen that Brown placed it in a new family: Ankylosauridae. Today, we regard ankylosaurids as one of several groups within Ankylosauria, but when Brown was writing, *Ankylosaurus* and other ankylosaurs were seen as part of Stegosauria. Throughout Brown's 1908 paper, *Ankylosaurus* is therefore referred to as a stegosaur. This classification persisted until 1923, when Henry Osborn proposed that ankylosaurs deserve to be recognized as their own distinct group of armor-plated herbivores.

Brown realized that *Stereocephalus* was also an ankylosaurid, and he suggested that it might be ancestral to *Ankylosaurus*. In 1910, Lambe

renamed *Stereocephalus* as *Euoplocephalus*, because *Stereocephalus* had already been given to an insect. Today, paleontologists think *Euoplocephalus* and *Ankylosaurus* are close relatives.

Ankylosaurus reconstructed

By combining all of the new anatomical details he had noted, Brown was able to produce a skeletal reconstruction (*see* above). Like so many early reconstructions, it became highly influential, and pictures of live ankylosaurids based on this diagram were still being published as recently as the 1980s.

Several details are now known to be substantially inaccurate, however. Since Brown thought that ankylosaurids were stegosaurs, he used parts of the stegosaur skeleton to fill in those regions missing from *Ankylosaurus*. He showed *Ankylosaurus* with the particularly robust forelimbs of a stegosaur, and with a pelvis in which both the ilium and pubis sport long, slender, forward-projecting prongs. He also reconstructed a short, drooping tail that ends with a blunt tip. Later discoveries substantially changed this view (*see* pp.78–79).

Left In Barnum Brown's original reconstruction, shown here, *Ankylosaurus* was assumed to be short-tailed, with a humped back and stegosaur-like hip girdle.

> ### DISCOVERY PROFILE
>
> | | |
> |---|---|
> | **Name** | *Ankylosaurus magniventris* |
> | **Discovered** | Montana, U.S.A., by Peter Kaisen, 1906 |
> | **Described** | By Barnum Brown, 1908 |
> | **Importance** | The first good ankylosaurid specimen to be discovered |
> | **Classification** | Ornithischia, Thyreophora, Ankylosauria, Ankylosauridae |

59

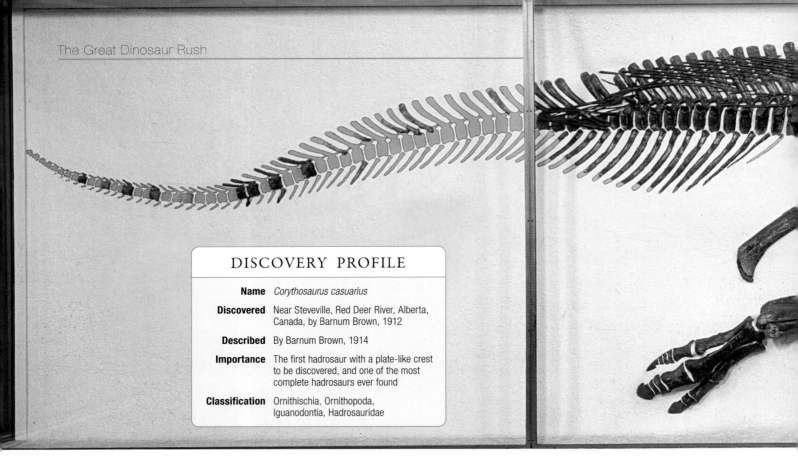

Barnum Brown's hadrosaurs

By 1913, duckbilled dinosaurs (hadrosaurs) were well known from many excellent skulls and skeletons, and the first crested hadrosaurs, such as spike-crested *Saurolophus* and hook-nosed *Gryposaurus*, had been described. The next stage in the understanding of hadrosaurs was the discovery that their crests were sometimes flamboyant and bizarre.

Above One of the first crested hadrosaurs to be discovered, *Hypacrosaurus* had a tall, rounded, plate-like crest.

Barnum Brown named a new crested hadrosaur in 1913, and the following year he described a second. Both have similar, plate-like crests, often likened to those of cassowaries (large, flightless birds native to Australasia). The first species, *Hypacrosaurus*, was initially known only from vertebrae, ribs, limb girdles, and limb bones, but these are notable for their huge size. Brown stated that *Hypacrosaurus* approached *Tyrannosaurus* in bipedal rearing height—

Hypacrosaurus means "nearly-the-highest lizard." The skull of *Hypacrosaurus* was unknown until 1924, when a second specimen was described.

Brown's other new hadrosaur, *Corythosaurus*, is far better known. It was named for a remarkable specimen discovered near Steveville on the Red Deer River, Alberta, in 1912. The specimen was almost entirely complete (only its forelimbs and part of the tail were missing), fully articulated, and preserved in a posture that made it look as if it had just lain on its side and died.

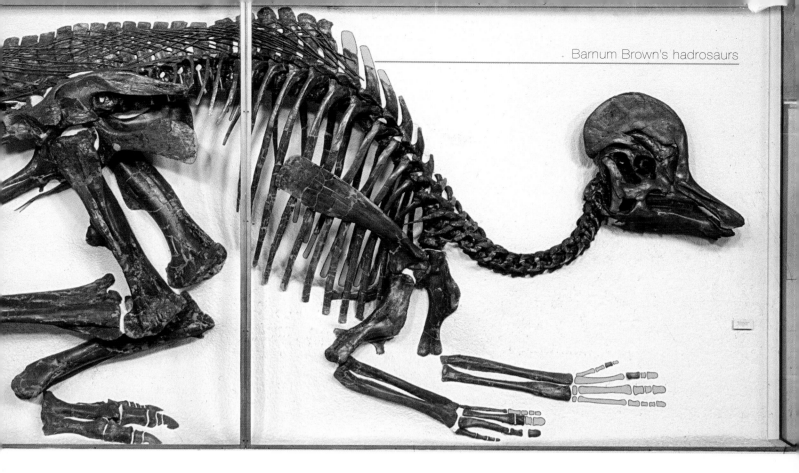

Corythosaurus retained a soft-tissue outline of the entire animal and detailed patches of skin were preserved. There was a ribbon-like crest of skin running along its neck, back, and tail; similar skin crests found in some other hadrosaurs have serrated or segmented margins.

Folded skin

Brown drew attention to parallel vertical skin folds in *Corythosaurus*. These were prominent in the shoulder region, and also visible across the thorax and trunk. He concluded that these must have been "present in life as loose folds of skin."

Brown thought that the specimen had vertical furrows in the skin next to the tall bony vertebral spines, so the reconstructions of live corythosaurs produced for his paper by Richard Deckert have a peculiar wrinkled, corrugated look to their bodies. There were large, conical scales arranged in widely spaced rows along the belly, and Brown described these as "limpet-like." They were separated by smaller, tubercle-like scales. He also mentioned fleshy pads on the feet.

This remarkable array of features made *Corythosaurus* one of the best known, if not *the* best known, hadrosaur in terms of both skeletal features and soft-tissue anatomy. Few hadrosaurs of this quality have been discovered since.

Posture and lifestyle

Brown's new *Corythosaurus* was preserved in a horizontal posture, in common with other articulated hadrosaur skeletons. Deckert illustrated the species swimming with its body horizontal, but showed it standing bipedally on land, with the tail contacting the ground. This "traditional" posture had been widely adopted ever since Dollo's work on *Iguanodon* and Leidy's study of *Hadrosaurus*. It was maintained by Brown and his colleagues because they thought that hadrosaur forelimbs indicated a bipedal lifestyle.

During the early 1900s, hadrosaurs were interpreted as amphibious animals that spent most of their time in water, although they were thought capable of walking and foraging on land. Brown thought that the deep, laterally compressed tail of *Corythosaurus* indicated swimming habits, and he also noted that the fossils associated with hadrosaurs were consistent with an amphibious lifestyle. This view persisted for decades to come. Brown and his colleagues also assumed that the plate-like crest of *Corythosaurus* was a narrow, cockscomb-like structure. However, during the 1920s Lawrence Lambe showed that the crests are complex and contain looping internal tubes.

Above Collected by Barnum Brown in 1912, this *Corythosaurus* was one of his most spectacular discoveries. Only its tail and hands are incomplete.

61

Alberta's spike lizard

Styracosaurus is one of the most striking and familiar horned dinosaurs. It was described by Lawrence Lambe in 1913 and was immediately recognized as a bizarre species of ceratopsian. Charles H. Sternberg found the fossil in what is now Dinosaur Provincial Park in Alberta. Although *Styracosaurus* was similar to forms such as the frilled *Triceratops*, it nevertheless was spectacularly new—and all the more so because the Sternberg specimen was the first complete horned dinosaur skull discovered in Canada.

Above Seen from the front, the long spikes on the frill of *Styracosaurus* must have given this dinosaur a striking, imposing appearance.

The deep-snouted skull of *Styracosaurus* sports a short frill, the rear edges of which are decorated with six long spikes that project backward and sideways. Smaller, spur-like spikes are present on the side edges of the frill. The skull is over 6 ft (2 m) long when the spikes are included, but without them the specimen is of more modest proportions, with a 20-in (50-cm) frill and a 30-in (75-cm) skull. There is also a long, straight nasal horn that projects slightly forward, but it broke before collection and its tip is missing from the original specimen. Lambe thought that the horn snapped about midway along its length, and a reconstructed tip added to the damaged horn produced an impressive length of nearly 24 in (60 cm). We know from new specimens, however, that the nasal horn was much shorter and blunter than Lambe thought: a more realistic length would be 12 in (30 cm). Old reconstructions of *Styracosaurus* therefore exaggerate the length of the horn.

Mistaken identity

When interpreting *Styracosaurus*, Lambe made another error,

A "BRISTLING" REPTILE

The descriptive terms that Lambe applied to *Styracosaurus* were as elaborate as its appearance. Although its name translates as "spike lizard," the Greek word *styrax* does not simply mean "spike" but relates specifically to a blade that projects from the side of a spear. In referring to the animal's spiky frill, Lambe wrote that the spikes "must have made this bristling reptile in life a veritable moving *chevaux de frise*." He apparently meant "*cheval de frise*," which is a wooden anticavalry device made of logs covered with spikes.

Right Shown here with the over-long nasal horn long thought correct, *Styracosaurus* was a rhinoceros-sized ceratopsian that weighed less than 2.2 tons (2 tonnes).

but one that is very understandable. The animal has oval depressions immediately above its eye sockets, which he assumed housed small horns. Referring to them as "incipient" horns, he suggested that they had become detached and lost during decomposition. This idea was accepted by other experts and was still being repeated as recently as 1990. We now know that pits like these are present in quite a few centrosaurines (the ceratopsid group that includes *Styracosaurus*). By studying the different growth stages of centrosaurines, scientists have shown that juveniles had small horns above the eyes, but these disappear by adulthood, leaving behind pits in mature centrosaurines. Recent work has shown that ceratopsian skulls changed substantially as the animals matured.

Another horned dinosaur

Styracosaurus was Lambe's second new horned dinosaur: he had named *Centrosaurus* in 1904, and he then named *Chasmosaurus* in 1914 and *Eoceratops* in 1915 (the last of these is no longer recognized as distinct). However, despite the distinctive appearance of *Centrosaurus*, doubts have sometimes been raised about the validity of this dinosaur. Some experts have suggested that *Styracosaurus* might have been

the male of *Centrosaurus apertus* from the same formation, and others have proposed that *Centrosaurus* and *Styracosaurus* are similar enough to belong in the same genus. Neither of these theories is currently regarded as correct because the two are different in many anatomical details and appear to belong to distinct lineages.

More specimens

The skull of *Styracosaurus* has often been featured in books and scientific articles. However, it is not well known that the lower jaws and most of the rest of the skeleton of the same original individual were also discovered, although these fossils were not collected from the field until 1935.

Many styracosaur specimens are known today, but little information has been published on the skeletons to date. A second *Styracosaurus* species, *S. ovatus*, was named in 1930, and a third species, *S. parksi*, followed in 1937. A revision of the styracosaurs published in 2007 showed that *S. ovatus* has a unique configuration of the spikes at the back of the frill and hence should be recognized as a distinct species (the spikes nearest the midline point inward in *S. ovatus*, whereas in *S. albertensis* they point outward). *S. parksi* is not distinct, but based on an *S. albertensis* specimen.

Above *Styracosaurus* was discovered in one of the richest dinosaur-bearing sites in the world: Dinosaur Provincial Park in Alberta, Canada. Hundreds of specimens continue to be collected from sites within the park, such as these hadrosaurs discovered in 1994.

Styracosaurus lived alongside a diverse assortment of herbivorous and carnivorous neighbors. In this reconstruction, *Styracosaurus*, with the large single horn jutting from its face, stands alongside *Chasmosaurus*, another ceratopsian. The hadrosaur *Brachylophosaurus* wanders by, and the small pachycephalosaur *Stegoceras* and the ankylosaur *Edmontonia* go about their business. However, one of the largest predators of the time, the tyrannosaur *Gorgosaurus*, has arrived on the scene.

The sauropods of Tendaguru

In 1906, German mining engineer Bernhard Sattler came upon what became one of the most famous dinosaur-bearing fossil sites in the world: Tendaguru in Deutsch Ostafrika (now Tanzania). The rich fossil assemblage at Tendaguru was soon brought to the attention of scientists. Between 1909 and 1913, the Museum für Naturkunde in Berlin led a series of expeditions to Tendaguru, during which it excavated and exported many dinosaur fossils.

Above The spectacular skull of the Tendaguru brachiosaur is one of the world's most remarkable sauropod fossils.

Without doubt the most spectacular discovery was the huge sauropod *Brachiosaurus brancai*, named in 1914 by Werner Janensch. Janensch concluded that this animal was closely related to a sauropod that American paleontologist Elmer Riggs had described from Colorado in 1903, *Brachiosaurus altithorax*.

Reconstructing a brachiosaur

Janensch and his colleagues in Berlin were working between the wars, at the height of a severe economic recession. They labored away to mount the Tendaguru brachiosaur and the other dinosaurs. The original plan was to make replicas of all the brachiosaur bones, and to mount *them* instead of the real things. This idea was abandoned, although the near-complete skull was deemed too valuable to be perched high in the air and a replica was put in its place. The real skull is very large, at nearly 3 ft (1 m) long, but it belongs to a smaller individual than that represented by the mounted skeleton. With its broad muzzle and arching nasal crest, it became the archetypal brachiosaur skull shown in life restorations.

Sauropods were generally regarded as amphibious for most of the 20th century, and thus it was suggested that the nasal crest of *Brachiosaurus* helped to keep the nostrils high up on the skull. This, it was thought, might have assisted the submerged animal when it brought its head to the surface to take a breath. Riggs viewed *B. altithorax* differently: he thought it was a giraffe-like, terrestrial browser that fed from the tops of trees, but his interpretation was ignored.

Radical interpretations

Janensch described the Tendaguru brachiosaur in a series of papers published between the 1920s and 1960s. The task of mounting the whole skeleton was completed in 1937. In some ways, Janensch's views on sauropods were farsighted and modern. He became convinced, for example, that the large openings on the sides of the vertebrae are for air sacs, and that these dinosaurs has pneumatic bones like those

DISCOVERY PROFILE

Name	*Giraffatitan brancai*
Discovered	Tendaguru, Tanzania (then Deutsch Ostafrika), by the Museum für Naturkunde expedition of 1909–1912
Described	By Werner Janensch, 1914
Importance	The first substantially complete brachiosaur to be discovered
Classification	Saurischia, Sauropodomorpha, Sauropoda, Brachiosauridae

in modern birds. This idea was first suggested half a century earlier, but was largely forgotten about. Janensch also thought that sauropods carried their necks in erect postures.

However, Janensch also held views that are now known to be incorrect. He mounted the brachiosaur with sprawling forelimbs in which the elbows point outward, for instance, and also thought that the hind feet were plantigrade: that is, that the entire length of the foot was placed flat along the ground. Evidence discovered since has shown that sauropods had erect limbs, and that a large pad under the foot prevented the foot skeleton from touching the ground.

FORGOTTEN BRITISH FINDS

After World War I, Germany lost her colonies to the Allied powers as part of the peace settlement, and Deutsch Ostafrika was awarded to Britain. The British had clearly been keeping close tabs on the progress of German paleontologists at Tendaguru, for as early as 1919 the British Museum (Natural History) was leading expeditions to the region. Although these ventures resulted in new scientific discoveries, the results were never published.

Different proportions

In 1903, Riggs drew attention to the unusual proportions of *B. altithorax*. Its forelimbs are particularly long, its shoulders were higher than its hips, and its tail was proportionally short. The neck and skull were unknown, so it seems that few artists ever chose to depict *B. altithorax* as a living animal.

Since the new, African brachiosaur was far better represented than the American *B. altithorax*, most people forgot that the American species was the "original" *Brachiosaurus*. After the 1930s, virtually every illustration of *Brachiosaurus* depicts *B. brancai* rather than *B. altithorax*. The two animals are actually quite different: *B. altithorax* had a longer and deeper body, a longer tail, and more widely splayed forelimbs. It now seems that the Tendaguru brachiosaur was different enough from the American species to warrant its own name, *Giraffatitan*, or "giant giraffe."

Above The gigantic skeleton of the Tendaguru brachiosaur is over 72 ft (22 m) long and 36 ft (11 m) tall. It is seen here with its head out of shot and the American *Diplodocus* in the foreground.

Below Brachiosaurs were unusually proportioned to most sauropods. Their tails were short, and their arms were particularly long.

67

The "parrot lizard" *Psittacosaurus*

During the 1920s, expeditions from the American Museum of Natural History in New York discovered many new dinosaurs on the high plateaus of Mongolia. Among the first of these to be published were two ornithischians, described in 1923 by Henry Fairfield Osborn. He suggested that one of the dinosaurs was related to the armored ankylosaurs, and the other to *Iguanodon* and its relatives.

Above *Psittacosaurus* is one of the most primitive of the ceratopsians, or horned dinosaurs. It could probably walk both on its hind legs and on all four limbs.

The skull of the first specimen has a short, deep, narrow rostrum that forms a curving beak reminiscent of a parrot's, and so Osborn named the species *Psittacosaurus*, or "parrot lizard." He suggested that the toothless beak was used for cutting and crushing: he noted its similarity to that of a turtle and proposed that it was well suited for dealing with tough plants. Pointed horns, called epijugals, project sideways from the cheek regions.

Osborn thought that a series of rounded tubercles on the right side of the skull represent the displaced remains of armor for protecting the throat and head. Perhaps, he argued, this hints at a covering of armor plates similar to that in ankylosaurs and stegosaurs. However, such tubercles have not been discovered on other psittacosaur specimens, and in 1924 Osborn noted that they were probably geological artifacts. Recognizing *Psittacosaurus* as the first member of a new family (Psittacosauridae), he proposed that the animal might be related to the more completely armored ornithischians of the Upper Cretaceous.

DISCOVERY PROFILE

Name	*Psittacosaurus mongoliensis*
Discovered	Near Uskuk, Tsagan Nor Basin, Gobi Desert, Mongolia, by the American Museum of Natural History Central Asiatic Expedition, 1922
Described	By Henry Osborn, 1923
Importance	The first discovery of a basal ceratopsian
Classification	Ornithischia, Marginocephalia, Ceratopsia, Psittacosauridae

A second ornithischian?

In 1923, Osborn gave the name *Protiguanodon mongoliense* to the other small ornithischian. The specimen was a partial skeleton just over 4 ft (1.2 m) long. Osborn thought it was an ornithopod similar to forms such as *Hypsilophodon* and *Thescelosaurus*, yet sufficiently different to deserve its own subfamily, Protiguanodontinae, within his concept of Iguanodontidae (a group that essentially included all ornithopods except the hadrosaurs). *Protiguanodon* was clearly similar to *Psittacosaurus*. Indeed, Osborn later concluded that the former was simply a juvenile of the latter, and today there is no doubt that *Protiguanodon* is the same animal as *Psittacosaurus mongoliensis*.

In his 1924 paper, Osborn produced the first skeletal reconstruction of a psittacosaur. The animal's limb proportions led him to suggest that psittacosaurs walked bipedally and in a "semi-erect" posture, with the body held horizontally but the tail sloping down toward the ground. He also showed how the hand was peculiar in having the third finger longer than the preceding two.

Misidentified bone

The strangest aspect of Osborn's work on these dinosaurs is that he did not compare them with ceratopsians. This seems odd, because the laterally compressed, toothless beak and flaring epijugal

horns of *Psittacosaurus* are so reminiscent of those known for geologically younger, more advanced ceratopsians, such as the ceratopsids.

Osborn's omission can be explained in part by his incorrect interpretation of the psittacosaur skull: he identified a bone at the tip of the upper jaw as the premaxilla, whereas the real premaxilla is located further back and the bone at the jaw-tip is in fact most similar to a bone unique to ceratopsians, the rostral. This identification was not suggested until 1956, when Alfred Romer proposed it in his monumental *Osteology of the Reptiles*. Before Romer's suggestion, dinosaur scientists had followed Osborn in misinterpreting the psittacosaur rostral as the premaxilla.

Common ancestors

Despite this problem, other experts had drawn attention to the similarity that *Psittacosaurus* has with ceratopsians as early as 1925. William Gregory and Charles Mook, in their description of *Protoceratops* published in that year, suggested that it and other frilled ceratopsians share so many features with *Psittacosaurus* "that

the existence of an earlier common ancestral stock is virtually demonstrated, *Psittacosaurus* retaining much the greater number of primitive characters." In the following decades, similarities between psittacosaurs and ceratopsians were mentioned many times, but the classification of *Psittacosaurus* within the Ornithopoda remained. At last, in the 1970s, it was proposed that psittacosaurs should be included within Ceratopsia.

Psittacosaurus is one of the very best-known dinosaurs, thanks to the discovery of hundreds of specimens that include many juveniles representing all stages of growth. Numerous psittacosaur species are also now known, varying in skull shape, body size, and other details.

Above *Protoceratops* was a rather more advanced ceratopsian than *Psittacosaurus*. It had a long bony frill and two pairs of fang-like teeth near the front of the upper jaw.

Left Unlike *Psittacosaurus*, *Protoceratops* had a body and limbs suited for full-time quadrupedality. It was close to the ancestry of the ceratopsids, the group that includes *Triceratops* and its relatives.

A new world of Asian theropods

The Central Asiatic Expeditions of the American Museum of Natural History in New York, led by Roy Chapman Andrews and Henry Osborn, were a great success. Although they did not discover the ancestors of modern humans, which was Osborn's primary intention, the scientists did come across a huge treasure trove of ancient mammals, lizards, and dinosaurs, all of which were new. Three exciting theropods were found during the second expedition, in 1923, all of which Osborn described a year later.

Above The original *Velociraptor* skull was lightly built and narrow, and the upper surface of the snout was depressed. Although the animal was impressively bird-like, Osborn regarded it as a sort of megalosaur.

THE *VELOCIRAPTOR* RIDDLE

Despite *Velociraptor's* importance, scientists were for many years surprisingly vague about its appearance. Judging from the scant material published about dinosaurs in the mid-20th century, it seems that most experts of the period believed that *Velociraptor* was a nondescript theropod little different from *Compsognathus*, albeit somewhat larger. The true appearance of *Velociraptor* and its relatives—the dromaeosaurids—was not to be revealed until the late 1960s, when John Ostrom described *Deinonychus*.

The first of these dinosaurs, *Velociraptor*, was less than 6 ft (2 m) long, and Osborn described it as alert and swift-moving. As for what sort of theropod the species was, Osborn regarded it as part of Megalosauridae. This seems unusual to modern eyes and is an example of how greatly views on theropod diversity and evolution in the 1920s differed from modern views. Osborn noted that the skull and teeth of *Velociraptor* appear to suit the capture of small, swift-moving prey, and he suggested that the strongly curved, compressed claws were used to grasp prey during capture. His suggestion that the relatively long skull and indication of a wide gape allowed *Velociraptor* to swallow prey that was both still alive and proportionally large is of particular interest. No proper studies of gape size in *Velociraptor* have ever been done, so we remain unsure if this speculation has merit.

Other small, fast hunters

Velociraptor was not the only new small theropod to be discovered on this expedition. Two others, *Oviraptor* and *Saurornithoides*, were

Right *Velociraptor* and its relatives were thought for decades to have scaly skin like crocodilians or lizards, but recent finds have revealed that they were feathered like birds.

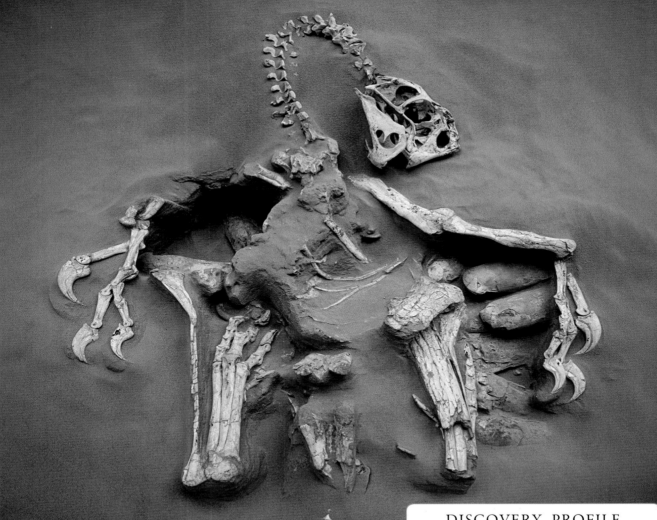

also found. In a move that today seems rather unimaginative, Osborn gave *Saurornithoides* and *Velociraptor* the same species name: *mongoliensis*.

Today, *Saurornithoides* is recognized as a troodontid: one of a group of bird-like theropods related to dromaeosaurids, but differing from them in having shorter arms and longer legs. Osborn initially compared the partially eroded *Saurornithoides* skull with that of toothed birds, but he later regarded it as another megalosaur-type theropod. He suggested that the animal's closely packed, coarsely serrated teeth indicated that it might be an egg eater. He did not elaborate further, and modern studies of troodontid teeth do not support this idea.

The third new Mongolian theropod was *Oviraptor*, a bizarre toothless animal. One of the most interesting details about its skeleton is that it preserves a wishbone, or furcula. Unfortunately, Osborn misidentified this as an "interclavicle." This could be why many paleontologists missed or ignored it in the years that followed.

Mysterious eggs

The *Oviraptor* specimen was found in association with eggs thought to belong to *Protoceratops*—itself a new discovery—and as a result *Oviraptor* was suspected of being a nest robber. This explains the dinosaur's species name *philoceratops*, which means "fond of ceratopsians." However, several associated skeletons and eggs belonging to *Citipati*, a close relative of *Oviraptor*, came to light in the 1990s. By this time scientists realized that the nest found with Osborn's Mongolian specimen belonged to the *Oviraptor* itself, and not to a *Protoceratops* after all.

Osborn's description of the three remarkable theropods was brief, but nothing else was forthcoming in the literature for decades to come. These remarkable animals were not revisited until the 1960s and 1970s, when they finally became the focus of detailed analysis.

DISCOVERY PROFILE

Name	*Velociraptor mongoliensis*
Discovered	Shabarakh Usu, Gobi Desert, Mongolia, by Paul Kaisen, in 1923
Described	By Henry Osborn, 1924
Importance	The first dromaeosaurid to be described
Classification	Saurischia, Theropoda, Coelurosauria, Dromaeosauridae

Above This specimen, dubbed "Big Mamma," shows *Citipati*—a close relative of *Oviraptor*—preserved on top of a nest full of eggs. Like birds, these dinosaurs sat on top of their eggs to protect and probably warm them. It now seems that the original 1923 *Oviraptor* specimen was not stealing *Protoceratops* eggs, but looking after eggs of its own.

The tube-crested hadrosaurs

Saurolophus and *Corythosaurus*, described in 1912 and 1914 respectively, showed that at least some hadrosaurs, the "duckbilled dinosaurs" of the Late Cretaceous, have unusual bony head crests. In *Corythosaurus*, this is a hollow structure with convoluted internal passages. But in 1920, an entirely different type of crest was discovered.

The impressive specimen was discovered by L. W. Dippell near Alberta's Red Deer River and described in 1922 by Canadian paleontologist William Parks. Its long crest is tubular and projects backward from above the eyes. Parks thought the dinosaur might be closely related to *Saurolophus*, which also has a backward-projecting bony crest. Accordingly, he named it *Parasaurolophus*, which means "near *Saurolophus*."

Parks noted, however, that the arrangement of the *Parasaurolophus* snout bones resembled that in *Corythosaurus*, and so he also suggested that this is where the affinities of *Parasaurolophus* might lie. Later work confirmed his theory.

Below This reconstructed *Parasaurolophus* skeleton belongs to the long-crested species *P. walkeri*. Not all *Parasaurolophus* species had long crests.

Strange tubes

A chance break through the crest enabled Parks to show that the structure incorporates four mysterious tubes divided by thin bony walls. Parks knew that more information could be obtained if he sawed the skull in half, but was "unwilling to injure the specimen so seriously."

During the 1920s, hadrosaurs were regarded as amphibious animals, and the hollow crest of *Parasaurolophus* was thought to be a specialization for such a lifestyle. In the 1930s, German paleontologist Martin Wilfarth even suggested that *Parasaurolophus* used its crest as a snorkel—however, the crest has no breathing hole at its tip. By the 1940s, Charles

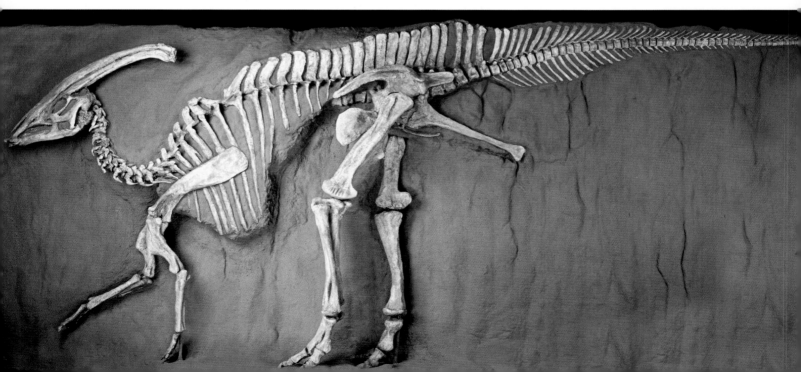

M. Sternberg was discussing the idea that the convoluted passages in the crest formed air-traps that prevented water from entering the lungs. The suggestion that hadrosaurs were amphibious has since been proved incorrect.

Parks also hinted at a display role for the crest. He thought that it was connected to a skin frill of the sort now known to have been present in other hadrosaurs. We know that artists were being advised to depict their crested hadrosaurs in this way because a contemporary painting produced for the Field Museum of Natural History by Charles Knight shows *Parasaurolophus* with a skin web connecting the crest to a frill that runs down the neck.

Vertebral feature

A peculiar feature of the first specimen of *Parasaurolophus* is its sixth dorsal vertebra, which has a disk-shaped expansion on its neural spine. Parks had already wondered about a soft tissue connection between the end of the crest and the animal's back, and so it was natural for him to propose that muscles and ligaments perhaps joined the crest's tip with this vertebra.

Parks also showed how *Parasaurolophus* has an unusual notch between two of its dorsal vertebrae. More recently, it has been suggested that this might have allowed the animal to "lock" its crest against its back. However, both the disk-shaped pad and the notch are the result of injury or disarticulation, and they were not normal features of this hadrosaur's skeleton.

DISCOVERY PROFILE

Name	*Parasaurolophus walkeri*
Discovered	Sand Creek, Red Deer River, Alberta, Canada, by L. W. Dippell, 1920
Described	By William Parks, 1922
Importance	Often regarded as the most bizarre crested hadrosaur, and the first tube-skulled hadrosaur to be discovered
Classification	Ornithischia, Ornithopoda, Iguanodontia, Hadrosauridae

Above Like all hadrosaurs, *Parasaurolophus* had hundreds of small, tightly packed teeth within its jaws. These formed rasping, pavement-like batteries.

Below Because the crests of hadrosaurs such as *Parasaurolophus* were almost certainly used in display, they were probably brightly colored, as shown here.

DIVERSE LIFESTYLES

It has often been stated that hadrosaurs are essentially all the same in the details of their skeletal anatomy, but this is not quite true: many subtle details hint at different specializations and ways of life. As William Parks noted, *Parasaurolophus* has particularly short, stout forelimbs, and its shoulder blade is huge for a hadrosaur. Its pelvis is also more robustly built than in some other hadrosaurs. The old idea that this species might have been the most aquatic hadrosaur—a view that came about because of its long, hollow crest—now seems incorrect. Its crest and other features instead suggest that it was strongly adapted for terrestrial life.

Dinosaurs of Transylvania

One of the most important contributions ever made to our understanding of European dinosaurs was the work of Hungarian nobleman and scientist Franz Baron Nopcsa (1877–1933). He is typically described as having an interesting and flamboyant personal life, and as the scientist who wrote about the unusual dwarf dinosaurs found in Romania's Late Cretaceous rocks. Less well known is that Nopcsa also discussed jaw function in dinosaurs, the origins of bird flight, regional geology, and theoretical tectonics.

Above A peculiar ornithopod unique to Romania, *Zalmoxes* was a medium-sized biped about 13 ft (4 m) long, with a broad, robust skull.

Much of Nopcsa's work was published during the so-called "quiet phase" in the history of paleontology, when comparatively little work was being done on dinosaurs. Nopcsa not only worked on dinosaurs from the Late Cretaceous rocks of Romania's Hateg Basin, he also discovered them himself, although his sister, Ilona, may have been the original finder of some of the fossils.

A new hadrosaur

In 1903, Nopcsa named a new hadrosaur from the Hateg Basin, calling it *Telmatosaurus transsylvanicus*. *Telmatosaurus* was a small hadrosaur, about 16 ft (5 m) long. Despite its late age, it is archaic compared to other members of the group. The dinosaur's remains were spectacularly complete, making it one of the best known Upper Cretaceous European dinosaurs. Despite this, the animal was mostly ignored by American hadrosaur experts and therefore remained little discussed until the 1990s, when its remains were properly analysed by David Weishampel and colleagues.

A little known ornithopod

Nopcsa's second Hateg Basin dinosaur, first named in 1902, was another ornithopod. This animal has had a complicated taxonomic history, but is currently known as *Zalmoxes*. Initially it was confused with a related ornithopod called *Rhabdodon*, and both kinds were thought to be *Iguanodon*-like animals. Later Nopcsa regarded them as relatives of

SEXUAL DIMORPHISM

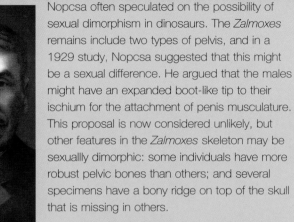

Nopcsa often speculated on the possibility of sexual dimorphism in dinosaurs. The *Zalmoxes* remains include two types of pelvis, and in a 1929 study, Nopcsa suggested that this might be a sexual difference. He argued that the males might have an expanded boot-like tip to their ischium for the attachment of penis musculature. This proposal is now considered unlikely, but other features in the *Zalmoxes* skeleton may be sexuallly dimorphic: some individuals have more robust pelvic bones than others; and several specimens have a bony ridge on top of the skull that is missing in others.

Hypsilophodon, or of *Camptosaurus* from the Morrison Formation. Few other paleontologists wrote about *Rhabdodon* during the 20th century, so it was never properly compared with other ornithopods. Recent work has shown that *Zalmoxes* and *Rhabdodon* belong together in a group, the Rhabdodontidae, which is positioned on the cladogram between *Hypsilophodon* and the iguanodontians (the group that includes *Camptosaurus*, *Iguanodon*, and the hadrosaurs).

The *Zalmoxes* remains that Nopcsa described were excellent, and enough is known about this dinosaur to allow an accurate reconstruction to be produced. It was a robust, round-bodied animal with a relatively large head and narrow jaw tips. Its limb proportions show that it was bipedal, but its broad body and hips suggest that its hind limbs were quite widely splayed compared to those of many other ornithopods.

More discoveries

Other ornithischians were found in the Hateg Basin. In 1915, Nopcsa named the ankylosaur *Struthiosaurus transylvanicus*. This dinosaur was originally named in 1871 for remains from Austria, and Nopcsa concluded that the Hateg Basin ankylosaur was similar to, and hence a close relative of, the Austrian species (*S. austriacus*). *Struthiosaurus* species have also been described from the Upper Cretaceous of France and Spain.

The Hateg Basin *Struthiosaurus* is known from skull material, a shoulder girdle, armor, and vertebrae. Its appearance in life remains unknown, but we do know that even adults of this dinosaur were small, having an estimated total length of about 10 ft (3 m). By 1923, the small size of the Transylvanian *Struthiosaurus* remains had led Nopcsa to suggest that these were dwarf, island-dwelling dinosaurs.

Hateg Basin's dinosaur assemblage also included sauropod remains, and again the animal concerned—now known as *Magyarosaurus*—was small compared to its relatives. At 16–20 ft (5–6 m) long, it was a relatively large animal, but for a sauropod it was tiny.

The "island" theory

During the 1920s, Nopcsa developed the idea that the Hateg Basin dinosaurs were unique, island-dwelling species, and that what is now the Hateg Basin was a Cretaceous island. He believed that the Hateg Basin dinosaurs were late-surviving descendants of species that were more widespread earlier during the Mesozoic era, and therefore that these animals had colonized the island at an early stage in the evolutionary histories of their respective groups. Paleontological work in the Hateg Basin effectively ceased after World War I. Scientists did not begin re-evaluating the unusual dinosaurs that Nopcsa had brought to the world's attention until the 1970s—and Nopsca's theories are widely accepted today.

Above Several species of *Struthiosaurus* inhabited Late Cretaceous Europe. The life appearance of this dinosaur is poorly known: compared to many other ankylosaurs, it was small and conservative.

The best-known prosauropod

Plateosaurus engelhardti was named in 1837 from fossils discovered in Heroldsberg, Bavaria, Germany, and described by Hermann von Meyer in the same year. He concluded that he was dealing with a gigantic animal related to *Iguanodon* and *Megalosaurus* and named it *Plateosaurus*, which means "broad lizard." It was not until the 1920s that German paleontologist Friedrich von Huene completed the surprising description of this animal.

Right Like all sauropodomorphs, *Plateosaurus* was long-necked. In contrast to most Triassic herbivores, it could therefore reach high up into the foliage.

Dinosaurs similar to, but smaller than, *Plateosaurus* were already known by the end of the 19th century. Little *Thecodontosaurus* was reported from the Triassic of England during the 1830s, and *Ammosaurus* and *Anchisaurus* were described from the Lower Jurassic of Connecticut during the 1880s and 1890s. By this time, scientists realized that these dinosaurs were related (all were seen as part of Theropoda), but *Plateosaurus* remained a mystery. A spectacular discovery changed this—and made it one of the best known Triassic dinosaurs.

Muddled remains

Additional species related to *Plateosaurus* were named in the first decade of the 20th century, and the taxonomy of these dinosaurs soon began to descend into chaos. In 1907 and 1908, Friedrich von Huene decided to redescribe the Heroldsberg plateosaur bones. Unfortunately, the plateosaur bones had become muddled with the teeth and skull bones of large predatory reptiles called rauisuchians, and of one rauisuchian in particular, called *Teratosaurus*. *Teratosaurus* has a deep, rectangular skull and massive recurved teeth. As a result, the plateosaur reconstructions produced by von Huene reveal that he wrongly imagined some of them to combine the broad-hipped, bulky body of a herbivore with the deep skull and recurved teeth of a voracious predator.

Preserved in mud

During the early 1920s, von Huene began working on excellent *Plateosaurus* specimens that had been discovered at Trossingen in southwest Germany. It was thought that these animals must have died after becoming stuck in deep mud. Previously, von Huene had concentrated on describing anatomy, interpreting phylogeny, and naming new species, but while working on the Trossingen fossils he turned his hand to taphonomy—the study of processes after death. He worked closely with an artist to reconstruct the Trossingen plateosaurs and their environment. It was his belief that the plateosaurs underwent annual migrations across a vast desert, and that small and weakened individuals had stopped at desert pools and become stuck.

Some of these skeletons were exquisitely preserved and still in fully articulated death poses. They were preserved as if they had slumped on their bellies, with their limbs folded at their sides and their necks curved round to the side. They included complete, three-dimensional skulls. The teeth at the tips of the jaws were conical and fang-like, whereas those further back were leaf-shaped and coarsely serrated, a combination suggesting that *Plateosaurus* was omnivorous. Its large, five-fingered hands sported large curved claws that might have been used in defense, in digging up roots, or in tearing at vegetation while feeding.

Bipedal or quadrupedal?

Von Huene's plateosaur reconstructions had always shown tail-dragging animals that held their bodies in a diagonal posture, but his work on the Trossingen material was different. Although the new reconstruction's body was upright and the tail's base dipped toward the ground, the tail was mainly parallel to the ground—although an awkward bend was built into the tail in order to get it into this posture.

Some authors have argued that *Plateosaurus* and its relatives walked quadrupedally, but von Huene thought that these dinosaurs were bipeds. Since the hands of these dinosaurs could not be rotated to get the palms to face the ground, some experts argue today that plateosaurs were incapable of quadrupedality.

In 1932, von Huene proposed that plateosaurs were closely allied with sauropods. He erected the new name Sauropodomorpha for this grouping, and suggested Prosauropoda for the group that included *Plateosaurus*. Although prosauropods may not have been the direct ancestors of sauropods, the idea that these Triassic animals were somehow involved in sauropod ancestry became well established.

Above *Plateosaurus* lived alongside large, dangerous theropods. Perhaps, as shown here, it defended itself by rearing up and striking out with its large thumb claws.

DISCOVERY PROFILE

Name	*Plateosaurus engelhardti*
Discovered	Heroldsberg in Bavaria, Germany, by Dr. Engelhardt, some time before 1837
Described	Fully, by Friedrich von Huene in 1926
Importance	The first large Triassic dinosaur to be discovered, and the first sauropodomorph to be discovered
Classification	Saurischia, Sauropodomorpha, Plateosauria

Cutler's articulated ankylosaur

When Barnum Brown described the first ankylosaurid skeleton in 1908 (*see* pp.58–59), he assumed that *Ankylosaurus* has a short, thoroughly unremarkable tail. Although his material showed that ankylosaurids are powerfully built, stocky dinosaurs with broad, armored bodies, little detailed information was available. However, by the 1920s, several new ankylosaurid specimens provided vital new information.

Below *Euoplocephalus* is younger and smaller than its close relative, *Ankylosaurus*. It was a broad-bodied, club-tailed dinosaur with horns at the back of its skull.

One of the most important ankylosaurid finds of this period was an incomplete specimen collected in 1920 from the Red Deer River in Alberta, Canada. Described by William Parks in 1924, it consisted of the hindquarters and tail, and it showed that ankylosaurids have a remarkably broad pelvis quite unlike the "stegosaur" pelvis reconstructed for *Ankylosaurus* by Brown. It preserved the armor in its life position and revealed that large, pointed scutes project in lines from the sides and upper surface of the tail. Even more remarkable was the fact that the vertebrae at the tail tip form a rigidly connected, stiffened handle-like structure, tipped with a massive bony club formed from coalesced scutes. Ankylosaurids, then, most certainly did not have the short, nondescript tails

that Brown had assumed. Parks thought that this ankylosaurid was a new animal, and named it *Dyoplosaurus acutosquameus*.

Bankside discovery

Far more complete than the *Dyoplosaurus* specimen, however, was an extraordinary articulated skeleton, preserved with intact armor, discovered by William Cutler, again near the Red Deer River. The skeleton was discovered in 1915. It was first mentioned in print in a 1926 newspaper article and was described by Hungarian scientist Franz Baron Nopcsa in 1928. Precisely where Cutler located the specimen has proved one of the great mysteries of Cretaceous dinosaur research. All we know is that he found the skeleton resting on its back in an awkward location halfway

up a steep cliff, and its excavation must have involved tremendous effort. The specimen was bought by the then British Museum (Natural History), and is still on display in London.

The "thorn lizard"

Referring to Cutler's specimen as "the finest armored dinosaur ever discovered and the single one in which nearly all parts of the dermal armor are preserved *in situ*," Nopcsa thought that it was worthy of its own species. He called it *Scolosaurus cutleri*, meaning "Cutler's thorn lizard." Its body is phenomenally broad, and its back virtually flat: whether this incredible breadth is natural has proved controversial. Much of the skeleton is complete, but the tail tip is missing. Nopcsa was unsure about this, and wondered whether *Scolosaurus* has a tail club like that of *Dyoplosaurus*. He also considered the possibility that the tail might finish abruptly "as in some of the fossil South American sloths."

The specimen's limb bones are short and very stout, and, based on the shape of the limb sockets and the articular surfaces of the limb bones, Nopcsa argued that both the humerus and femur must project outward and sideways from the body. He therefore described the animal as walking with its feet spread far apart. A reconstruction produced for *The Illustrated London News* by Alice B. Woodward, and used by Nopcsa in his technical paper, depicts *Scolosaurus* as a tail-dragging, turtle-like creature with short, widely spread limbs obscured from view by the wide body. This view of ankylosaurids is now known to be incorrect—their limbs did not sprawl and they did not drag their tails—but it remained popular for the next few decades.

How many species?

The status of *Scolosaurus* is something about which ankylosaur experts still argue. Some are confident that it is a specimen of *Euoplocephalus*, and that the only features that make Cutler's ankylosaurid seem unusual are the result of erosion, damage, or distortion. Others have argued that the pattern of armor on the specimen is different from that of *Euoplocephalus*, and that *Scolosaurus* and *Dyoplosaurus* differ from *Euoplocephalus* in having tall, laterally compressed, conical spikes located halfway along the tail. Here, we regard all three dinosaurs as the same species.

Above This model *Euoplocephalus* shows the wide beak, deep snout, and large, triangular horns seen in this dinosaur. A mobile bony plate was embedded in the eyelid.

Below The gigantic tail club of ankylosaurids was almost certainly used as a weapon.

DISCOVERY PROFILE

Name	*Euoplocephalus tutus*
Discovered	Dead Lodge Canyon, banks of the Red Deer River, Alberta, Canada, by William E. Cutler, in 1915
Described	By Lawrence Lambe, in 1902, with Cutler's specimen described by Franz Baron Nopcsa in 1928
Importance	The first fully articulated ankylosaurid to be discovered
Classification	Ornithischia, Thyreophora, Ankylosauria, Ankylosauridae

79

China's first dinosaur

In terms of dinosaurian heritage, China is known today to be one of the richest countries in the world. But dinosaurs were not scientifically discovered in China until surprisingly recently. The first Chinese dinosaur to be described and named was the amazingly long-necked sauropod *Euhelopus*.

Right *Euhelopus* was a particularly long-necked sauropod. Experts continue to argue about the life appearance of these dinosaurs, but a posture with an erect neck, such as that shown here, is likely.

The remains of *Euhelopos* were gathered during a Sino-Swedish expedition in Shandong, China, in 1922. The collection included the hadrosaur *Tanius*, some possible stegosaur and theropod bones, and two partial sauropod skeletons. The collection was sent to Swedish paleontologist Carl Wiman of the University of Uppsala, and it was he who described it in 1929.

"Marsh foot"

The Shandong fossils were collected by Austrian paleontologist Otto Zdansky, and Wiman honored him by naming the Lower Cretaceous sauropod *Helopus zdanskyi*, which means "Zdansky's marsh foot." The name *Helopus* had already been given to a seabird in 1832, so a new name—*Euhelopus*, or "true marsh foot"—was eventually published by Alfred Romer in 1956.

Intact skull

Euhelopus was unlike any known sauropod, and Wiman decided it belonged to a new family, dubbed the Helopodidae (later changed to Euhelopodidae). It is interesting for several reasons, not least its beautifully preserved, three-dimensional skull. Sauropod skulls are rare, partly because they are fragile and easily destroyed by predators or erosion; by the 1920s they had been described for only a few species, such as *Camarasaurus* from the Morrison Formation.

THE LONGEST NECK

Even by sauropod standards, the neck of *Euhelopus* is very impressive. When Wiman described the Shandong specimen, it was the longest-necked sauropod then known: he noted that the neck boasts 17 long, narrow vertebrae. Based on the posture of the articulated fossil, he thought that *Euhelopus* kept its head raised in life, and his reconstruction shows the animal with its neck lifted up at an angle of about 40°.

In *Euhelopus*, the snout is broad and rounded and the teeth are spatulate (broad and rounded at the ends). As with so many sauropods, the nostril openings are located high up on the snout and close to the eyes, a feature that Wiman interpreted as allowing the animal to breathe at the surface of the water without needing to raise the whole of its head.

Foot shape

Wiman's chose *Euhelopus* for the generic name of the new sauropod because he believed the hind foot was well suited for supporting the animal's weight on soft mud (the shape of the fore foot was not known). He even compared the broad shape of the hind foot with snowshoes, and illustrated the specialist footwear in his 1929 paper. The inside digits of the dinosaur's

DISCOVERY PROFILE

Name	*Euhelopus zdanskyi*
Discovered	Ningxiachou, Shandong, China, by Otto Zdansky, Sino-Swedish expedition, 1922
Described	By Carl Wiman, 1929
Importance	The first Chinese dinosaur to be named, and the longest-necked sauropod then known (in terms of number of vertebrae)
Classification	Saurischia, Sauropodomorpha, Sauropoda, Euhelopodidae

Orbit, or eye socket

Infratemporal fenestra

Mandible, or lower jaw

Naris, or nasal opening

Antorbital fenestra

Maxilla, or upper jaw

Tooth

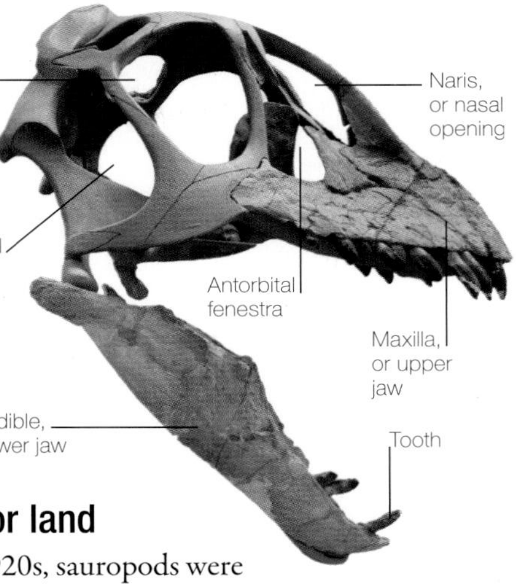

Left *Euhelopus* had a short-snouted skull. Its teeth are broad, rounded at the ends, and heavily worn, and were almost certainly used to crop leaves from high up in trees.

foot have three claws, and Wiman showed two of these as being particularly broad and thus efficient at spreading weight, comparing them to the splayed hooves of marsh-dwelling antelopes.

However, these claws look so broad in Wiman's description because he showed them as lying on their sides. Correctly oriented, they are deep and narrow. In fact, we know from other specimens that sauropod feet were not broad and splayed as Wiman thought, but compact and proportionally small for the size of the animal. Wiman also thought that *Euhelopus* might be four-toed, but this is almost certainly not the case and the fifth digit simply appears to be missing.

Water or land

By the 1920s, sauropods were generally regarded as amphibious animals of lakes and swamps, mostly thanks to the earlier ideas of Owen, Cope, and Marsh. This was despite the fact that Riggs argued for terrestrial habits as early as 1904. Wiman thought that the supposedly snowshoe-like feet of *Euhelopus* were compatible with an aquatic lifestyle, and that its long neck allowed it to breathe from the surface while standing in deep water. His article includes a reconstruction of two *Euhelopus* individuals behaving this way. When feeding in deeper water, he imagined that these sauropods might be able to rear into a bipedal posture in order to reach the surface, and he illustrated this behavior as well.

81

Are birds dinosaurs?

The idea that birds are theropod dinosaurs, closely related to dromaeosaurs, such as *Deinonychus*, is widely accepted. This theory was once based entirely on skeletal evidence, but the discovery of numerous feathered dromaeosaurs and other theropods shows that complex feathers were not unique to birds, but appeared earlier in theropod evolution.

In fact, primitive birds are very similar to the small, primitive members of other, closely related theropod groups. This implies that birds are fundamentally similar to other dinosaurs and are merely one theropod group among many.

Above New discoveries have shown that small dromaeosaurs, such as *Microraptor* shown here, were more bird-like than anyone ever thought possible.

Right Many modern birds have claws on their hands, but these are usually hidden by feathers and are typically vestigial and without a function. This claw belongs to an emu.

Far right Outstanding fossils from Liaoning Province in China, such as this *Microraptor*, preserve spectucular fossil feathers.

Above Known among paleontologists as AMNH 7224, this fossil is one of the best known of the Ghost Ranch *Coelophysis* specimens. The small bones visible within its rib cage were once thought to be evidence of cannibalism.

The Ghost Ranch theropods

Ever since the discovery of *Compsognathus* in the late 1800s, paleontologists knew that theropods included small as well as large species. However, the small theropods of the Triassic and Early Jurassic were known only from fragmentary remains. The discovery of hundreds of articulated small theropod skeletons in Late Triassic rocks at Ghost Ranch, New Mexico, was, therefore, a major breakthrough.

The Ghost Ranch *Coelophysis* fossils were discovered by accident. In 1947, Edwin Colbert and his colleagues from the American Museum of Natural History traveled to New Mexico in search of fossils. One colleague, George Whitaker, found some bone fragments that Colbert identified as *Coelophysis*. The team began a thorough search and soon came upon a rock layer that contained a mass of bones—the area eventually yielded hundreds of complete and near-complete skeletons. One of the team declared it to be "The greatest find ever made in the Triassic of North America." The bones were confined to an area measuring 20 ft by 65 ft (6 m by 20 m), and included individuals of different ages.

Colbert was confident that the theropods were *Coelophysis bauri*, named by Edward Cope in 1889. But whereas the Ghost Ranch skeletons were complete, Cope's material was scrappy. The Ghost Ranch skeletons showed that *Coelophysis* has a lightly constructed skull, short forelimbs, and a long, slender body.

Evidence of cannibalism

Some of the Ghost Ranch theropods appeared to contain the remains of juvenile members of their own species in their stomach contents. Too large to be embryos, they were interpreted by Colbert as evidence for cannibalism. Cannibalism is common in predatory reptiles, so its presence in theropods was not surprising. However, the alleged baby *Coelophysis* specimens were eventually shown to be small crocodilians.

Mysterious mass death

The Ghost Ranch mystery is why so many individuals were preserved together. It seems that the animals were rapidly buried and not left exposed to scavengers or on the surface, where moving water and sediments would have helped

break up the skeletons. Colbert wondered if poisonous gases from a volcano might have killed them. Today, we still do not know: Colbert's speculation from the 1940s has yet to be replaced by a compelling new hypothesis. Whatever killed the dinosaurs, they had formed a large group beforehand. Perhaps they were attracted by an abundant food source, such as spawning fish.

The archetypal theropod

Due to the hundreds of specimens, *Coelophysis* became the archetypal small theropod. Despite its fame, however, a lengthy description of *Coelophysis* was not published until 1989, and even then, doubts remained about whether the Ghost Ranch theropods were the same animal as Cope's original *Coelophysis*.

The last centrosaurine

Ceratopsids, bristling with horns and spikes and sporting huge bony frills, were well known for being flamboyant and bizarre, but *Pachyrhinosaurus canadensis*, described and named in 1950, was something truly new.

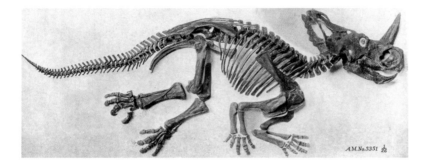

Above Ceratopsids typically had prominent nasal horns. *Centrosaurus*, shown here, was a typical member of the ceratopsid sub-group Centrosaurinae.

The geologist O. A. Erdman found fragments of the new dinosaur at Scabby Butte, Alberta, in 1945 and sent them to Charles M. Sternberg for further study. The following year, Sternberg collected skull fragments at the same site. However, much of the specimen had apparently been removed by local people who used the area as a picnic spot. A more complete skull was then discovered at Scabby Butte by R. Steiner, and it was this specimen that showed Sternberg that *Pachyrhinosaurus* is a highly unusual ceratopsid.

Battering ram nose?

Pachyrhinosaurus is among the largest of ceratopsians: the lower jaw of one of Sternberg's specimens is over 19 in (48 cm) long. The animal is also massively built, being very thick-boned, with fist-like lumps projecting from the sides of the skull roof. Instead of a nasal horn, *Pachyrhinosaurus* has a huge, flattened boss covering its snout. It is decorated with coarse striations and pits, so Sternberg thought that it might have functioned as a battering ram. He stated that it would have been covered in a horny sheath in life and thus was even larger than the boss appeared. Later work showed that the nasal boss is not really solid bone, but is far more porous. A "battering ram" function is therefore unlikely.

Racial senescence

Sternberg implied that the "great thickening of bone" in *Pachyrhinosaurus* was "suggestive of the freakish development that took place among some of the dinosaurs near the close of the Cretaceous." During the early 20th century, some scientists suggested that Cretaceous dinosaurs might have suffered from what was known as "racial senescence." According to this idea, certain groups of organisms somehow run out of genetic potential, and take to growing bizarre horns, crests, and growths, rather than evolving to deal with the problems placed before them.

When paleontologists discussed this theory in relation to dinosaurs, they had the elaborately crested lambeosaurs in mind, as well as the thick-skulled pachycephalosaurs and the horned dinosaurs. Sternberg's comment implies that he was also proposing that *Pachyrhinosaurus* might have been "racially senescent." The hypothesis of "racial senescence" never had any firm basis and is not considered scientifically valid today.

DISCOVERY PROFILE

Name *Pachyrhinosaurus canadensis*

Discovered Scabby Butte, Alberta, Canada, by O. A. Erdman and Charles M. Sternberg, 1945–1946

Described By Charles M. Sternberg, 1950

Importance Often regarded as the most bizarre horned dinosaur

Classification Ornithischia, Marginocephalia, Ceratopsia, Ceratopsidae

---

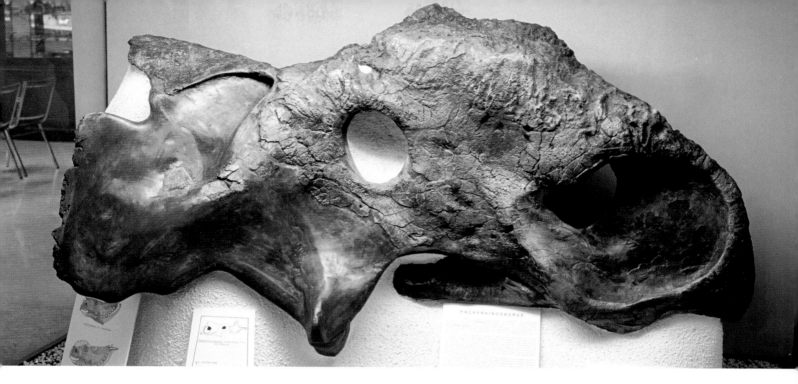

Sternberg's model

To depict the life appearance of this unusual dinosaur, Sternberg created a model of the head as he thought it had appeared in life. Its beak is narrower than the rest of the skull and its jaws are deep, short, and massive. He modeled it with a short, smooth-edged, rounded frill (only the base of the frill was preserved) and also showed the bony boss as a fairly neat, oval structure projecting upward from the top of the skull like a flat-topped tree trunk.

Sternberg's model is interesting because he gave the animal cheeks: he assumed that ceratopsians probably needed these to allow them to retain vegetation within the mouth while it was being sliced up by the teeth. It is clear that most of the life restorations of *Pachyrhinosaurus* produced in the following decades used this model head for inspiration.

A new family

Sternberg thought that *Pachyrhinosaurus* belonged to a totally distinct lineage of ceratopsians, which he called the

Pachyrhinosauridae. However, later work showed that *Pachyrhinosaurus* was a centrosaurine, and a close relative of *Styracosaurus*. In fact, it is now known that *Pachyrhinosaurus* is the last and the largest known member of the group and perhaps the most specialized.

With the naming of *Pachyrhinosaurus* in 1950, the description of new horned dinosaurs from North America effectively ceased. It was generally thought that the continent had no new ceratopsian species to yield. As we will see, this proved incorrect.

Above The remarkable near-complete skull of *Pachyrhinosaurus* was, when described in 1950, utterly different from that of the other horned dinosaurs known at the time.

Below Unlike *Centrosaurus* and most ceratopsians, *Pachyrhinosaurus* lacked a nasal horn. Instead, it had a thickened "nasal boss."

The Dinosaur Renaissance

Following the golden era of discovery during the early 20th century, scientific interest in dinosaurs waned mid-century and exploration largely ceased. This suddenly changed in the 1960s when John Ostrom announced the discovery of *Deinonychus*, an agile predator, similar in many details to the early bird *Archaeopteryx*. Ostrom wondered if *Deinonychus* and other dinosaurs were warm-blooded. Suddenly, dinosaurs became interesting again, and a new generation of scientists were inspired to study them. The dinosaur renaissance had begun.

Facing page In the 1960s, 70s, and 80s, new discoveries overturned the ideas of the early 20th century. Suddenly, dinosaurs began to be portrayed as active, agile, successful animals. Some widely held views remained incorrect, however: small theropods such as *Deinonychus*, shown here, were portrayed as scaly-skinned—we now know them to have been feathered.

Below The medium-sized theropod *Deinonychus*, announced in 1969, led many to realize that dinosaurs were not the dull, slow, stupid creatures of popular myth.

Montana
Navajo County, Arizona | Mesa County, Colorado

1960–1989

In the 1960s, dinosaur research emerged from the quiet phase of the 1940s and 50s in truly spectacular fashion. This invigorated paleontological movement is dubbed the "dinosaur renaissance." Several discoveries and proposals contributed to the sense of renewed scientific progress, the most important of which were those made by American paleontologist John Ostrom of the Peabody Museum of Natural History at Yale University. Some of Ostrom's earliest work focused on the duckbilled hadrosaurs: they were not amphibious animals as previously thought, but terrestrial browsers. Ostrom's ideas changed our understanding of hadrosaur biology.

Rio Grande do Sol
San Juan | La Rioja
Cerro Condor
Neuquen Province
La Colorado

1963 Triassic *Herrerasaurus* is named

1970 *Dilophosaurus* finally recognized as a new kind of theropod
1970 Mongolian "terrible hands" *Deinocheirus* is named

1972 Jensen reports new giant sauropods from Colorado

| 1960 | 1963 | 1966 | 1969 | 1972 | 1975 |

1962 "Different-toothed lizard" *Heterodontosaurus* is named

1964 Ostrom and colleagues discover *Deinonychus*

1969 *Deinonychus* is named

1974 New Mongolian pachycephalosaurs are described

Surrey, England 🔲

Dundgov 🔲
South Gobi Aimak 🔲 🔲 Amtgay
🔲 Nemegt basin

Wujiabai 🔲 🔲
🔲 Jianbei County
🔲 Dashanpu

Gadoufaoua 🔲

Muttaburra 🔲
🔲 Roma

Mapheteng 🔲
🔲 Herschel

Dinosaur Cove 🔲

O strom and his colleagues discovered a new theropod in the Cloverly Formation of Montana in 1964, and named it *Deinonychus* in 1969. It has a raised second toe that bore a large, sickle-shaped claw, and so walked on only two of its toes. Ostrom inferred from this that *Deinonychus* "must have been a fleet-footed, highly predaceous, extremely agile and very active animal," and that compared to modern reptiles it had "an unusually high metabolic rate." Ostrom also reexamined the oldest bird, *Archaeopteryx*. Many similarities were shared by *Archaeopteryx* and small theropods, and Ostrom argued that birds had evolved from *Deinonychus*-like theropods. In a series of articles published in the 1970s, he resurrected Huxley's idea of bird-dinosaur affinities, first proposed in the 1860s.

1976 More complete *Heterodontosaurus* is reported
1976 *Ouranosaurus* named from Niger

1980 *Minmi* is named from Australia
1980 A new "segnosaur," *Erlikosaurus*, is named

1983 *Baryonyx* discovered in England

1985 *Supersaurus* and "*Ultrasaurus*" are named by Jensen
1985 The horned abelisaurid *Carnotaurus* is named

1978 Ostrom publishes influential *National Geographic* article
1978 Dong and colleagues name *Yangchuanosaurus*

1979 The "segnosaurs" are recognized
1979 Horner and Makela publish on hadrosaur nests and babies
1979 Tiny babies of *Mussaurus* named from Argentina

1989 *Chasmosaurus mariscalensis* is named

1986 *Baryonyx* is named

Above Recent finds prove that *Deinonychus*, shown here attacking the herbivore *Tenontosaurus*, was feathered and more bird-like in appearance than even Ostrom had thought.

Another important figure in this phase of paleontological discovery was Robert Bakker, one of Ostrom's students at Yale. Inspired by *Deinonychus* and other recent finds, Bakker argued that the anatomy and evolutionary history of dinosaurs showed how they were not slow-moving, cold-blooded, evolutionary "dead-ends," as was generally assumed. Instead, he proposed, they were warm-blooded animals that led highly active, fast-paced lives. Furthermore, they were not dead-ends, but lived on in the form of birds and, far from being failures, were one of evolution's greatest success stories.

Superiority of the dinosaurs

During the late 1960s, Bakker went even further, claiming that dinosaurs were competitively superior to the ancestors of mammals. He suggested that direct competition between the two groups ultimately forced mammals to lead a furtive life in the shadows for the next 160 million years or so. Bakker also reexamined the sauropods, arguing in 1971 that they were fully terrestrial animals, best imagined as elephantine giraffes. The prevailing view since Victorian times was that sauropods had been amphibious swamp-dwellers.

A highly capable artist, Bakker was able to illustrate his articles with attractive, detailed drawings, and he portrayed dinosaurs in particularly provocative poses, showing them fighting, running, and jumping. His dinosaurs looked different from the stocky, shapeless, fat-bodied animals of earlier artists: they were svelte, muscular, and decorated with stripes and other dashing color schemes. He even reconstructed some theropods with feathery plumes or furry coverings, and showed land-bound sauropods rearing up on their back legs to battle one another for mating rights.

Fierce debate

Many scientists were not convinced by Bakker's arguments, which ignited a huge amount of interest and debate. This in turn inspired numerous articles for the public in journals such as *National Geographic*. The Bakker vision of sleek, hot-blooded dinosaurs was quickly and enthusiastically absorbed into popular culture, featuring in comics and highly successful books, notably Adrian Desmond's *The Hot-Blooded Dinosaurs*.

More exciting dinosaur discoveries from around the world further increased public and scientific interest in dinosaurs. New Jurassic dinosaurs from China, found during the late 1970s, made it clear that it had an outstanding Mesozoic fossil record. In the decades that followed, China became the most important country in the world in terms of influential dinosaur discoveries. Polish–Mongolian expeditions brought back amazing new dinosaurs from the Gobi Desert, and Africa and South America also revealed strange new species. The idea that dinosaurs were more complex animals than previously thought was reiterated in 1979, when Jack Horner and Bob Makela published evidence that hadrosaurs had practiced complex parental care. Studies such as this attracted more and more scientists to the study of dinosaurs.

Above American paleontologist John Ostrom, shown here with *Deinonychus*, produced carefully argued, detailed studies on dinosaur anatomy, biology, and behavior. More than any other single person, he revolutionized our view of dinosaurs.

Right This now iconic illustration of *Deinonychus*, produced by American paleontologist Robert Bakker, helped popularize the new view of dinosaurs as agile, successful animals.

93

The twin-crested theropod

Above The *Dilophosaurus* skull is highly distinctive. In addition to its fragile, plate-like crests, it has a shallow lower jaw and a prominent notch in the upper jaw's toothrow.

In 1942, Jesse Williams, a Navajo, discovered one of the 20th century's most newsworthy North American dinosaurs near Tuba City in northern Arizona. The skeleton was found in the rocks of the Kayenta Formation. Williams alerted a field party from the University of California Museum of Paleontology to the site, and they were able to recognize the skeletons of three theropods, two of which yielded good material.

Left The true function of the twinned crests present in *Dilophosaurus* is unknown. If they were used in display, they were probably brightly colored, as shown here.

Paleontologist Samuel Welles named the theropod as a new species of *Megalosaurus*: *M. wetherilli* in the brief, initial report of 1954. The species name honored John Wetherill, described by Welles as "explorer, friend of scientists, and trusted trader and counselor to the Navaho." *Megalosaurus wetherilli* is today known as *Dilophosaurus wetherilli* and its initial description as a species of *Megalosaurus* might appear odd to modern eyes.

Once *Megalosaurus* was described in England during the 1820s, it became standard practice to classify new theropod species within the same genus, even when they came from rocks that were substantially older, or substantially younger, than those that yielded the original *Megalosaurus*. The end result was that theropods from the Triassic, Jurassic, and Cretaceous, and from all around the world, were labeled as *Megalosaurus*. Today, it is understood that the name *Megalosaurus* should be restricted to the Middle Jurassic species *M. bucklandii*. However, *Megalosaurus* was still being used as a "taxonomic wastebasket" when Welles published his first, preliminary report.

How old?

During the late 1950s, the age of the Kayenta Formation became the subject of argument. A 1958 paper had argued for a Late Triassic age. Based on how "advanced" *M. wetherilli*

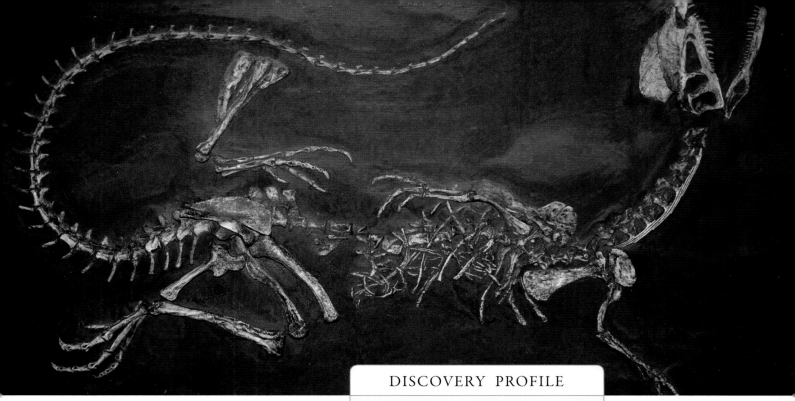

appeared to be compared to Triassic theropods such as *Coelophysis*, Welles concluded that the Kayenta was Early or Middle Jurassic. These differing conclusions inspired Welles to look anew at Kayenta geology in 1964, and it was while on this trip that another Kayenta theropod specimen was discovered.

This seemed to be a larger, more robust individual of *M. wetherilli*, preserved with an articulated skull. A plate-shaped structure is preserved on the skull's upper surface. This was initially thought to consist of bones that belong inside the skull, but had been pulled upward and outward by a scavenger. However, its position and shape showed that it must have been one of a pair of plate-shaped crests. This discovery caused Welles to reexamine the best of the three 1942 specimens. Sure enough, it preserves the bases of two crests that broke away before fossilization. It was obvious that the Kayenta theropod was not *Megalosaurus* but something entirely new, and in 1970 Welles named it *Dilophosaurus*, meaning "double-crested lizard."

The 1942 and 1964 theropods were at first assumed to be the same species, but Welles argued that the latter specimen was different, and in fact different enough to deserve its own new genus. Unfortunately, Welles died in 1997 and never got round to naming this second dinosaur, and his proposition has mostly been ignored and is generally forgotten today.

If he is right, then the exact size and shape of the crests in *D. wetherilli* remains a mystery, because the only complete crests yet discovered belong to the larger animal.

Bony headgear

Welles's view that the Kayenta Formation was Jurassic was vindicated—it is currently regarded as Early Jurassic—and *Dilophosaurus* remains one of the best-represented Early Jurassic theropods. Thanks to its discovery, scientists knew that some theropods have flamboyant bony headgear. But what did the animals do with these structures? It has been suggested that crests like this functioned in social and sexual display. However, this remains nothing more than a speculation, with proper hypothesis testing yet to be performed. In the meantime, finds from Antarctica, China, and elsewhere have shown that members of several theropod lineages have crests even more remarkable than the twin plates of *Dilophosaurus*.

Above *Dilophosaurus* was a lightly built, slender theropod about 20 ft (6 m) long. It was not known that the original specimen, shown here, was crested until a second specimen was found in 1964.

China's missing Jurassic dinosaurs

Above
Yangchuanosaurus,
one of China's
best-preserved
large theropods,
was initially identified as a
megalosaur, but is more
closely related to *Allosaurus.*

By the 1970s, Chinese dinosaurs remained mysterious and poorly known. When the first complete remains were finally discovered, they were novel and highly distinct. These new finds allowed China at last to join the "dinosaur renaissance," and it has since contributed in spectacular fashion.

DISCOVERY PROFILE

Name	*Yangchuanosaurus shangyouensis*
Discovered	Near Shangyou Reservoir, Sichaun Province, China, by workmen, 1977
Described	By Dong Zhiming, Zhang Yihong, Li Xuanmin, and Zhou Shiwu, 1978
Importance	China's most spectacular big theropod, and one of the country's first well-preserved theropod skeletons
Classification	Saurischia, Theropoda, Allosauroidea, Sinraptoridae

Right One of
China's best known
Jurassic sauropods
is *Mamenchisaurus.*
Its neck contains
19 vertebrae and,
in the largest species,
is more than 10 m
(33 ft) long, making it
one of the longest necks
of any known animal.

Among the most impressive Chinese dinosaurs were those discovered in Yongchuan County, Sichaun Province, in an Upper Jurassic rock unit known as the Upper Shaximiao Formation. One of the best-preserved Yongchuan dinosaurs was a new large theropod discovered in 1977 by workmen renovating the Shangyou Reservoir. They realized the fossil's importance and alerted the staff of the Chungking Museum of Natural History. In 1978, the Shangyou theropod was described by Chinese paleontologist Dong Zhiming and colleagues. They named it *Yangchuanosaurus shangyouensis*, meaning "Yongchuan lizard from Shangyou."

Classic death pose

The *Yangchuanosaurus* skeleton was 26 ft (8 m) long, spectacularly complete, and well preserved. It lay on its side with its neck and skull curved backward in a classic death pose, its tail curved upward, and its hind limbs folded beneath its body (its arms had been lost before discovery). Dong and colleagues surmised that the dinosaur might have become trapped in mud when chasing prey. However, its death posture is typical for dinosaurs and the specimen was not preserved in a manner consistent with miring: animals that die after getting caught in mud typically have their legs stuck vertically in the substrate.

Like so many other theropods described in previous decades, *Yangchuanosaurus* was originally identified as a megalosaur. Later work showed that it belonged to a new group (Sinraptoridae) closely related

Left This replica of the original *Yangchuanosaurus* skeleton shows how the predator is preserved in the classic "opisthotonic" death posture, in which the neck is curved up and over the back.

to *Allosaurus*. Its skull is deep and short-snouted, with tall, narrow bony openings on the sides and paired bony ridges running along the upper surface of the snout and skull roof. The tall neural spines on its vertebrae must have formed a narrow ridge along its back. *Yangchuanosaurus* was clearly a large, powerful animal, and like other big, deep-skulled theropods, it was certainly a predator of other large dinosaurs.

Unique features

One of the most interesting things about *Yangchuanosaurus* is that it was distinct from other large theropods found elsewhere—or so it was thought at the time. This helped to confirm suspicions raised earlier by the discovery of *Mamenchisaurus* and the other Chinese dinosaurs (*see* box). It seemed that, during the Middle and Late Jurassic, eastern Asia was an island continent home to unique dinosaurs. This isolation appears to have persisted until the Early Cretaceous, when land bridges enabled animals from Europe to reach eastern Asia, and vice versa.

Although most of China's Jurassic dinosaurs remain unique, newer work has shown that groups such as the sinraptorids were probably widely distributed before the isolation of eastern Asia. Jurassic Europe shared sinraptorids with China, for example, and some European Jurassic sauropods appear to have been close relatives of Chinese forms such as *Mamenchisaurus*.

The Lower Shaximiao Formation

Upper Shaximiao dinosaurs such as *Yangchuanosaurus* were spectacular, but there was an even richer assemblage in the geologically older Lower Shaximiao Formation that underlay it. Here, Middle Jurassic theropods and primitive stegosaurs were found alongside archaic sauropods and small bipedal ornithischians.

HISTORY OF PALEONTOLOGY IN CHINA

Knowledge of Chinese dinosaurs has developed relatively slowly. Following the description of the notably long-necked sauropod *Euhelopus* in 1929 (*see* pp.80–81), together with the hadrosaur *Tanius* and several other specimens, a few new kinds were announced over the years. Another particularly long-necked sauropod, *Omeisaurus*, was reported in 1939, and a large theropod, *Szechuanosaurus*, was named in 1942. A most unusual hadrosaur, *Tsintaosaurus*, was named from Laiyang in 1950, and four years later a sauropod with a neck even longer than that of *Euhelopus*, called *Mamenchisaurus*, was described from Sichuan. However, these discoveries only hinted at the riches still to come from Chinese rocks.

The best Lower Shaximiao site, Dashanpu, preserved so many dinosaur skeletons that an open-air museum was built there.

Further spectacular finds over the last 30 years have established China as the most important dinosaur-bearing country in the world. Dozens of new species belonging to entirely new dinosaur groups have revealed whole new chapters in our understanding of dinosaur evolution and biology.

"Terrible hand" and the fighting dinosaurs

Polish and Mongolian paleontologists published an exciting assortment of new Cretaceous dinosaurs from Mongolia during the 1970s. These fossils were collected on the Polish–Mongolian expeditions of 1963–65, 1967–69, and 1970–71, and they provided a huge amount of new information on dinosaur diversity and evolution.

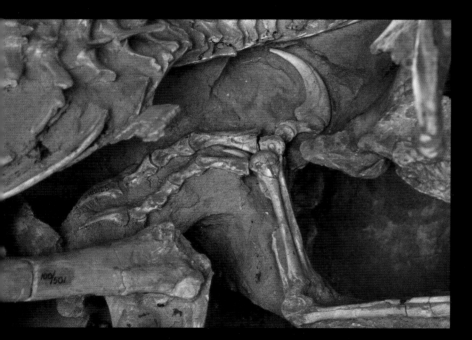

Above The famous "fighting dinosaurs" truly were locked in combat at the moment of death. The *Velociraptor's* sickle-shaped raised foot claw, shown here, was jammed right up against the herbivore's neck.

Among the new Mongolian dinosaurs was an astounding theropod known from its shoulder girdles, long arms, and gigantic three-clawed hands (ribs and other fragments were discovered too). The animal was described by Polish paleontologists Halszka Osmólska and Ewa Roniewicz in 1970, who named it *Deinocheirus mirificus*—the "unusual terrible hand."

Deinocheirus was certainly a large dinosaur: Osmólska and Roniewicz assumed that it was *Tyrannosaurus*-sized, in which case it was a formidable beast indeed. Its arms are almost 8 ft (2.5 m) long and its hands are broad, but its claws are short, weakly curved, and blunt compared to those of many other theropods.

Unusual theropod

Although *Deinocheirus* was clearly a theropod, nothing else like it was known. It was therefore given its own new family, Deinocheiridae, and its large size and relatively thick-walled bones led Osmólska and Roniewicz to regard it as a carnosaur rather than a coelurosaur. Due to its supposedly primitive hand structure, there was speculation that deinocheirids evolved during the Jurassic from megalosaur-like ancestors. The great length and slender proportions of the arm bones made this unlikely, however, and in fact the arms were similar to those of ostrich dinosaurs such as *Ornithomimus*.

Since the 1980s, paleontologists have tended to accept that *Deinocheirus* is in fact a gigantic ostrich dinosaur, though it is still so different from the other ostrich dinosaurs that it must have belonged to a distinct lineage within the group. If this is the case, *Deinocheirus* was almost certainly not a superpredator, but a herbivore or omnivore, and its claws were more likely better at manipulating foliage than disembowelling other dinosaurs.

Tantalizing clues

In their 1970 article on *Deinocheirus*, Osmólska and Roniewicz mentioned some other fossils worthy of comparison. They referred to work

DISCOVERY PROFILE

Name	*Deinocheirus mirificus*
Discovered	Altan Ula III locality, Nemegt Basin, Gobi Desert, Mongolia, by Zofia Kielan-Jaworowska, 1965
Described	By Halszka Osmólska and Ewa Roniewicz, 1970
Importance	One of the most bizarre and enigmatic dinosaurs
Classification	Saurischia, Theropoda, Coelurosauria, Ornithomimosauria

by Russian paleontologist E. A. Maleev, who in 1954 described some gigantic claws and other remains from the Nemegt Formation; he concluded that they belonged to a new turtle-like reptile, which he named *Therizinosaurus*. Osmólska and Roniewicz countered that these claws were more likely to be of theropod identity, and further discoveries proved them right (*see* pp.110–111). They also mentioned a gigantic hand claw from Gadoufaoua, Niger, and argued that this too was from a theropod.

These tantalizing fossils suggested that giant big-clawed theropods were widespread during the Cretaceous, but it was not clear whether all of these animals were deinocheirids, or if they belonged to distinct groups that were only distantly related.

Locked in combat

The 1970–71 expedition made another amazing theropod discovery: a spectacularly complete *Velociraptor* was preserved locked in combat with a specimen of the primitive ceratopsian *Protoceratops* (*see* pp.178–179). The pair became known as the "fighting dinosaurs."

Why did the animals die fighting? Theories include that both were buried by a collapsing sand-dune, that they were smothered by a sandstorm, and that they sank into soft sediment at the edge of a lake. It is likely that both animals inflicted fatal wounds on the other and that they perished locked together.

Left The gigantic arms of *Deinocheirus* probably belonged to a huge ostrich dinosaur, not to a super-predator. But without more data, we can only guess how this dinosaur made a living.

HAND FUNCTION

Osmólska and Roniewicz noted that the broad hands and weakly curved claws of *Deinocheirus* were not well suited for grasping, but concluded that the arms' long reach might have made *Deinocheirus* skilled at "tearing dead or weakly agile prey asunder." Roughened areas and bony pits and scars indicated that the animal had sustained various injuries to its hands during its life. The idea that the hands might have functioned in predation was repeated by many other authors.

The good mother lizard

A discovery in Montana brought about a complete overhaul of ideas about the parenting strategies of dinosaurs. By studing newly discovered eggs, nests, and babies belonging to the new duckbilled hadrosaur *Maiasaura*, paleontologist Jack Horner showed that dinosaurs provided extensive parental care. They built nests, and they protected, and perhaps even fed, their offspring.

Above
Compared to its more flamboyant relatives, *Maiasaura* was a dull, conservative hadrosaur.

Below Years of excavation at the maiasaur nesting grounds in Montana have provided a wealth of information on the behavior of these dinosaurs.

Before the 1970s, most paleontologists thought that female dinosaurs left their babies to fend for themselves once they had buried their eggs, as do turtles. This may have been true of some dinosaurs, but the diversity of parenting styles seen in living reptiles strongly suggests that many dinosaurs were better parents.

Horner's discoveries were made by chance in 1978 at a Montana rock and fossil shop. Horner was shown several tiny bones by the shop owner,

Marion Brandvold, and immediately realized that these were the bones of baby duckbilled dinosaurs, or hadrosaurs. They came from the rocks of the Two Medicine Formation near Choteau, Montana, and the fact that Horner had also discovered a fossil egg in the Two Medicine Formation indicated that these rocks might be very special indeed.

Horner and his colleague Bob Makela began excavating a piece of land owned by the James and John Peebles families, from where the baby

dinosaurs had come. They were lucky enough to uncover what appeared to be the remains of a preserved, bowl-shaped hadrosaur nest that contained the remains of 15 juveniles. This was strong evidence of post-hatching parental care.

Parental care

It seemed to Horner and Makela that the babies stayed in their nest after hatching and were looked after by their parents until large enough to leave. Although the fully formed wrist and ankle bones indicated that the young were quite capable of activity outside the nest, the presence within the nest of small eggshell fragments showed that the babies remained there for an extended period (during which time they trampled the eggshell). Wear on the babies' teeth and the presence of fossilized plant material suggested that the babies were fed by the parents.

The discovery of several other nests in the vicinity suggested that hadrosaurs formed nesting colonies, with the nests separated by a distance equivalent to that of an adult hadrosaur. By gathering together in this way, hadrosaurs might have relied on group defense and on a sort of communal early-warning system to remain safe from predators. By 1982, Horner was able to show that the sedimentary layers containing the nests were built on top of older layers that also contained nests: this demonstrated that the animals returned to the same nesting site repeatedly. They were practicing "site fidelity," a form of behavior also seen in modern birds that nest in colonies.

It was obvious that these hadrosaurs represented a new dinosaur. In honor of their newly discovered parenting skills, Horner and Makela named the species *Maiasaura peeblesorum*, meaning "good mother lizard of the Peebles families." *Maiasaura* appears to be a rather conservative hadrosaur, but it does have one unusual feature: a solid ridge-like crest, triangular in cross-section, present above the eyes of adults.

Conflicting interpretations

In the late 1980s, Horner and colleague David Weishampel suggested that, rather than being able to walk around after hatching, baby hadrosaurs had weak limbs and were essentially nest-bound. This idea was challenged in 1996, when it was argued that the hips and limbs of hadrosaur hatchlings were actually fully ossified, as Horner and Makela stated back in 1979. However, subsequent discoveries made elsewhere in North America provided support for the hypothesis that hadrosaurs nested in colonies, and that they also practiced post-hatching parental care.

Left Evidence indicates that parent maiasaurs fed their babies in the nest, and that the babies stayed in the nest for an extended period.

<table>
<tr><td colspan="2">DISCOVERY PROFILE</td></tr>
<tr><td>Name</td><td>*Maiasaura peeblesorum*</td></tr>
<tr><td>Discovered</td><td>Choteau, Montana, U.S.A., by Jesse Williams, *c.*1978 (adult material collected by Jack Horner and Robert Makela)</td></tr>
<tr><td>Described</td><td>By Jack Horner and Robert Makela, 1979</td></tr>
<tr><td>Importance</td><td>First hadrosaur shown to nest in colonies and practice post-hatching care of babies</td></tr>
<tr><td>Classification</td><td>Ornithischia, Ornithopoda, Iguanodontia, Hadrosauridae</td></tr>
</table>

The behavioral traits
true of maiasaurs
were almost certainly
true of all hadrosaurs.
Brachylophosaurus,
a close relative of
Maiasaura (*see* previous
page), probably also
nested in colonies and
looked after its babies.
Here, *Brachylophosaurus
canadensis* bellows a
call, warning her herd
of an ambush by a
Daspletosaurus torosus
(background).

Thick-headed dinosaurs

Above
Pachycephalosaurus was the largest of the boneheads and reached 16 ft (5 m) in length. Its thick, rounded skull differed notably from the flat skulls of forms such as *Homalocephale*.

Below *Homalocephale*, like all boneheads, had short arms, a broad body, and a stiff, narrow tail. Its skull roof was flat and wide. Lacking armor or other defensive weapons, boneheads probably relied on speed to escape predators.

The pachycephalosaurs, sometimes called the boneheads or domeheads, are bipedal Cretaceous ornithischians with close links to the ceratopsians. They are one of the most poorly known groups of dinosaurs and have had a confused history.

The first good pachycephalosaur skull, from the North American species *Stegoceras validum*, was described in 1924. Unfortunately, because its teeth are similar to those of the theropod *Troodon*, *Stegoceras* and *Troodon* then became regarded as the same animal. As a result, pachycephalosaurs were referred to as "troodontids" during the middle decades of the 20th century.

In 1945, Charles M. Sternberg realized that the original *Troodon* tooth was from a theropod, and that *Stegoceras* and its giant cousin,

Pachycephalosaurus (named in 1943), were not troodontids at all but deserved their own group. He named the group Pachycephalosauridae and included it within the Ornithopoda. Recent work has removed the pachycephalosaurs from among the ornithopods; the most important discoveries allowing this reclassification were some remarkable new pachycephalosaurs published by Polish paleontologists Teresa Maryańska and Halszka Osmólska in 1974.

Three new species

During their report of the discoveries made during the Polish–Mongolian paleontological expeditions of the late 1960s and early 1970s (*see* pp.98–99), Maryańska and Osmólska named three new pachycephalosaurs. All of them were represented by good skull material, and all came from the Upper Cretaceous of the Gobi Desert.

DISCOVERY PROFILE

Name	*Homalocephale calathocercos*
Discovered	Nemegt, Nemegt Basin, Gobi Desert, Mongolia, by Zofia Kielan-Jaworowska, 1965
Described	By Teresa Maryanska and Halszka Osmólska, 1974
Importance	First Asian pachycephalosaur known from good skeletal remains; first flat-headed pachycephalosaur
Classification	Ornithischia, Marginocephalia, Pachycephalosauria

Of the new species, *Tylocephale gilmorei* was known only from its skull and lower jaw. *Prenocephale prenes* included a skull as well as pelvic bones, thigh bones, and parts of the tail. But by far the best represented of the trio was *Homalocephale calathocercos*: in addition to a skull and vertebrae, its remains included pelvic and hind limb bones.

Strange features

Most pachycephalosaurs have domed skulls, whereas the roof of *Homalocephale*'s skull is flat. Rounded openings are present at the back of the skull roof (these are a normal feature of dinosaur skulls, but in dome-skulled pachycephalosaurs have closed over). The skull is also decorated with lumps, pits, and nodes. Although the front of the snout is missing, it is assumed (based on other pachycephalosaurs) that *Homalocephale* has delicate, narrow jaw tips and fang-like teeth at the front of the upper jaw.

The hips of *Homalocephale* are broad and unusually long ribs grow from the sides of the vertebrae at the base of the tail. Tendons form a tightly interwoven basket-like sheath around the vertebrae in the end half of its tail.

Biology and lifestyle

Maryańska and Osmólska went on to make a number of speculations about pachycephalosaur biology and lifestyle. The large eye sockets of their *Homalocephale* specimen indicated good vision; the structure of several pachycephalosaur skulls even suggested that the animals' eyes faced partly forward, giving some degree of binocular vision. The rounded edge of the premaxillary bones at the tip of the upper jaw led Maryańska and Osmólska to propose that pachycephalosaurs lack beaks, as is typical of ornithischians, but have fleshy lips instead. This idea has not been supported by other experts, however, and the presence of preserved beak tissue in some ornithischian fossils shows that their conjecture was false.

Head-banging speculation

Maryańska and Osmólska made several other claims about pachycephalosaur biology. They thought that these dinosaurs might have used their stiff tails as props when resting, that the wide hips allowed live birth, and, most famously, that the thick-boned domed or flattened skull roofs were used as weapons, either against predators or among individuals of the same species during the breeding season. Later work rejected or falsified these proposals, but the only one that could be subjected to proper analysis was the "head-banging" idea.

The hypothesis was supported by Briton Peter Galton in 1970, and it rapidly became popular. Soon, head-banging pachycephalosaurs were depicted in virtually every dinosaur book. However, an analysis of the interior of the pachycephalosaur skull has recently raised serious questions about the head-banging hypothesis (*see* pp.176–177).

Above *Prenocephale*'s dome-shaped head was decorated with lines of small bony tubercles. Like virtually all pachycephalosaurs, it was a medium-sized dinosaur less than 10 ft (3 m) long.

Below The incomplete skull of *Homalocephale*, shown here, has a thickened roof, and small bony lumps decorate its surface. These features are typical of pachycephalosaurs.

The different-toothed dinosaurs

The heterodontosaurids, or "different-toothed lizards," are among the most confusing and enigmatic of all ornithischians. Paleontologists first saw how bizarre these dinosaurs are when the 1961–62 British–South African Expedition to South Africa and Basutoland (modern Lesotho) came upon a strange skull.

The skull was immediately assigned to a new species, *Heterodontosaurus tucki*, by expedition members A. W. "Fuzz" Crompton and Alan Charig. The rocks that yielded the skull were thought at the time to be Late Triassic in age but are now regarded as Early Jurassic.

Crompton and Charig described how the skull has three different types of teeth. At the tip of its upper jaw, the teeth are incisor-like. Further back are large, canine-like fangs, and further back still is a series of column-shaped chewing teeth with chisel-like crowns. Having different tooth types is a condition known as heterodonty. It is associated with mammals

and not usually with reptiles, and it suggests that the heterodontosaurids were using different teeth for different tasks.

As is usual for ornithischians (bird-hipped dinosaurs), the tip of the *Heterodontosaurus* lower jaw is formed from a toothless predentary bone, and in life both this and the tip of the upper jaw were covered by horny beak tissue. Hollowed areas on the sides of the jaws suggest that heterodontosaurids had cheeks, allowing food to be retained in the mouth while chewing. Heterodontosaurids belong to the universally herbivorous ornithischians, and Crompton and Charig assumed that they were herbivores, given their chisel-like cheek teeth.

Left Though long assumed to be herbivores, the large fangs and long, grasping hands of heterodontosaurids suggest the possibility that they were omnivores.

DISCOVERY PROFILE

Name	*Heterodontosaurus tucki*
Discovered	Herschel, Cape Province, South Africa, by members of British–South African Expedition to South Africa and Basutoland, 1961–62
Described	By A. W. Crompton and Alan Charig, 1962 (first description of complete skeleton, 1976)
Importance	First well-preserved heterodontosaurid
Classification	Ornithischia, Heterodontosauridae

Fast-running biped

In 1976, Albert Santa Luca and his colleagues announced details of a near-complete *Heterodontosaurus* skeleton. Discovered in 1966, it has long arms with large muscle-attachment sites, long, five-fingered hands, and long lower leg bones. This evidence suggested to Luca that heterodontosaurids were fast-running, bipedal animals that used their hands to grasp and tear vegetation. A reconstruction produced to accompany the 1976 paper showed *Heterodontosaurus* standing in the kangaroo-like posture typical of previous decades, in which the body is oriented diagonally and the tail droops to the ground.

In their influential 1974 paper on dinosaur monophyly, Robert Bakker and Peter Galton illustrated parts of the specimen and argued that its hand and ankle anatomy are "virtually identical" to that of Triassic sauropodomorphs and theropods. If true, this offered compelling support for the newly resurrected idea that dinosaurs formed a clade—a group of organisms that share a single ancestor (*see* pp.10–11).

Early origins

The discovery of *Heterodontosaurus* gave a new insight into the early evolution of ornithischians. Although it has been proposed that heterodontosaurids might be ornithopods or close relatives of marginocephalians, the most recent phylogenetic studies concluded that they are one of the most basal clades

within Ornithischia. If this is correct, then the long, grasping hands of these animals might be primitive features retained from the dinosaur common ancestor.

Few early ornithischians have been discovered since *Heterodontosaurus*. *Eocursor* from South Africa and *Pisanosaurus* from Argentina were described from the Upper Triassic, and *Lesothosaurus*, *Stormbergia*, and the basal thyreophoran *Emausaurus* were described from the Lower Jurassic. Furthermore, many fossils once identified as Triassic ornithischians have been reidentified as crurotarsans or silesaurids, or as indeterminate relatives of archosaurs.

It now seems that ornithischians were rare during the earliest stages of dinosaur history, with only a handful of lineages present. Why ornithischians remained so scarce until well into the Jurassic is a mystery. One suggestion is that they were able to diversify only once extinction events at the end of the Triassic removed competing groups of non-dinosaurian herbivores.

MYSTERIOUS LARGE CANINES

We have no concrete data about the diet of heterodontosaurids, but their robust skulls and sharp fangs suggest that they may have eaten at least some animal material. *Heterodontosaurus* certainly has substantial canine-like teeth, more accurately called caniniforms. These are very similar to the fang-like teeth seen in a number of small herbivorous mammals alive today, such as musk and muntjac deer. The male deer use their teeth in territorial fights, so it is possible that heterodontosaurids may have done likewise.

The spectacular sailback

By the early 1960s, Cretaceous dinosaurs from Africa were poorly known. Ernst Stromer brought a collection of fossils to Germany in 1911, but these were destroyed during World War II. And French scientists found remains of North African dinosaurs during the 1950s and 60s, although these provided little information. Then, in 1965, another Frenchman discovered one of Africa's most famous dinosaurs, *Ouranosaurus nigeriensis*.

The discoverer of *Ouranosaurus* was Philippe Taquet from the Museum National d'Histoire Naturelle in Paris, who accepted an invitation to explore Gadoufaoua in Niger to prospect for fossils. In the Tamachek language of the Tuareg people, "Gadoufaoua" literally means "the place where the dromedaries are afraid to descend among the bumpy rocks," but despite this reputation, the location proved to be an important place for dinosaur finds.

The *Ouranosaurus* skeleton is well represented and partially articulated. It includes over 70 vertebrae, most of the pectoral girdle and forelimbs, pelvic girdle, and hind limbs, and much of the skull. A second, almost complete skeleton was discovered later, and many additional bones from the species were also found later in the same deposits.

Spiny vertebrae

Taquet named his dinosaur after the sand monitor, a large lizard found in the area and which is known as *ourane* in Arabic. This name in turn means "brave" or "fearless." *Ouranosaurus* was clearly a close relative of *Iguanodon* and other iguanodontian ornithopods, but Taquet found its evolutionary affinities difficult to determine. Its most remarkable feature is the very tall, laterally compressed neural spines on its dorsal vertebrae. These are superficially similar to the tall-spined vertebrae known in *Spinosaurus*, and Taquet assumed that they supported a tall "sail" that ran along the animal's back.

The function of these structures remains an interesting and still unanswered aspect of dinosaur biology. One popular idea has been that they had a key role in temperature control, and that the animals used them to absorb warmth from the sun or to cool down when overheating. However, in 1997 it was questioned whether these dinosaurs even had sails: their robust, rectangular bony spines most resemble similar structures found in bison, and perhaps show that these dinosaurs had muscular humps rather than sails.

Duckbill

Ouranosaurus, in contrast to *Iguanodon*, has a flaring, duckbilled muzzle. Its bony nostril openings are located some distance away from the edge of the snout and are surrounded by enormous concave areas on the snout's upper surfaces. *Ouranosaurus* shares these skull features with hadrosaurs. Taquet clearly illustrated and described the skull in 1976, but the

DISCOVERY PROFILE

Name	*Ouranosaurus nigeriensis*
Discovered	Gadoufaoua, Niger; by Philippe Taquet, 1965
Described	By Philippe Taquet, 1976
Importance	One of the best-known African dinosaurs
Classification	Ornithischia, Ornithopoda, Iguanodontia

duckbill feature was nevertheless overlooked by some of the artists and scientists who illustrated and wrote about *Ouranosaurus* during the 1970s and 1980s. Some assumed that it had a narrower, deeper snout, rather like that of *Iguanodon*.

Debate over posture

Taquet believed that *Ouranosaurus* had a relatively short tail compared to those of other iguanodontians, and he also thought it was a biped that walked with the body held diagonally. He thought that it was perhaps in the habit of standing quadrupedally when feeding from the ground, however. Views on the posture of these dinosaurs changed after the 1980s, when new work suggested that they often walked quadrupedally and with the body held horizontally.

The hadrosaur-like skull of *Ouranosaurus* proved important in a debate over hadrosaur ancestry that took place during the 1980s. Taquet noted that *Ouranosaurus* sufficiently resembled hadrosaurs to be involved in their ancestry. In 1990, hadrosaur expert Jack Horner took this further: he argued that *Ouranosaurus* shared a number of anatomical features with lambeosaurine hadrosaurs; *Ouranosaurus* and the lambeosaurines should thus be grouped together in a clade termed Lambeosauria. The other hadrosaurs—the hadrosaurines—were held to be more closely related to *Iguanodon*. Hadrosaurs do not form a clade if this theory is correct, but arose at different times from separate ancestors. The view was not supported by later analyses, but arguments over the evolutionary relationships of *Ouranosaurus* continued.

NASAL BUMPS

Much less well known than its duckbill is the fact that *Ouranosaurus* also has tall, rounded bumps on the roof of its skull, located above and in front of the eyes. Philippe Taquet called these "nasal protuberances," and their function is unknown. It is possible that they might have been used in head-shoving matches, or perhaps they were decorations that played a part in courtship displays.

The remarkable therizinosaurs

By the 1970s, paleontologists thought that representatives of all the major dinosaurian groups were known. Although unfamiliar species and genera continued to be found, the possibility of entirely new families was considered unlikely. The announcement of a group of bizarre, hitherto unknown dinosaurs from the Upper Cretaceous of Mongolia—the segnosaurs—was therefore a surprise. The first of these, discovered in 1973 during the Joint Soviet–Mongolian Paleontological Expedition, was named *Segnosaurus galbinensis*.

DISCOVERY PROFILE

Name	*Segnosaurus galbinensis*
Discovered	Amtgay, southeast Gobi Desert, Mongolia, by members of Joint Soviet–Mongolian Paleontological Expedition, 1973
Described	By Altengerel Perle, 1979
Importance	First therizinosauroid known from good remains
Classification	Saurischia, Theropoda, Coelurosauria, Therizinosauroidea

Right We now think that, like other bird-like theropods, *Segnosaurus* and its relatives were feathered. With their massive hand claws, broad bodies, and long necks, they must have made an unusual sight.

Named by Mongolian paleontologist Altangerel Perle in 1979, *Segnosaurus* is known from a reasonable amount of material, although only a single fragment of its skull survives. Its slender lower jaw has an unusual, down-curved tip and houses small, leaf-shaped teeth. Its arm bones preserve evidence of powerful musculature and the only known hand claw is strongly curved and flattened from side to side. The pelvis is amazingly broad, flaring outward and showing that the animal had a tremendously broad abdomen. Both the pubis and ischium are directed backward—this is the opisthopubic condition present in both ornithischians and birds.

The stout hind limb bones of *Segnosaurus* look best suited for a heavy, slow-moving animal. Its feet are strange: unlike those of a theropod, in which the first toe is strongly reduced and no longer has a contact with the ankle joint, the first toe in *Segnosaurus* appears to contact the ankle joint and to touch the ground. The metatarsal bones are short, and it is their shortness that enables the first toe to reach ground level. The large foot claws are curved and flattened from side to side.

Perle thought that segnosaurs were highly unusual theropods, but pointed out that this classification was provisional. He noted that *Segnosaurus* seemed to be quite different from *Deinocheirus* (*see* pp.98–99), and also that it

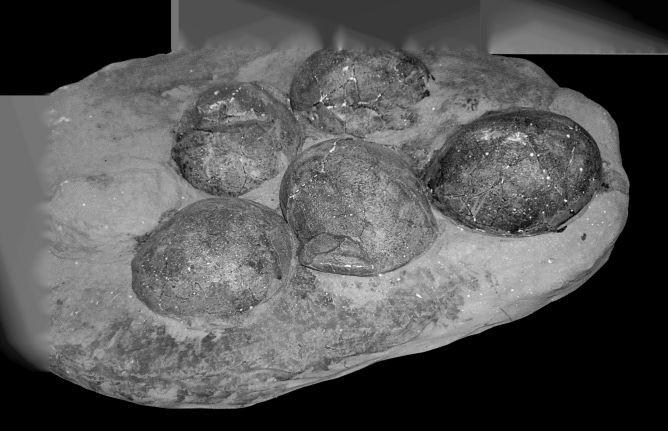

differs in the shape of its arm bones and in the length of its claws from *Therizinosaurus cheloniformis*, a poorly known, long-clawed Mongolian fossil initially described in 1954 as a gigantic, turtle-like reptile. Russian herpetologist Anatole Rozhdestvensky showed in 1970 that *Therizinosaurus* was not turtle-like, but was in fact a gigantic, long-armed theropod.

Another segnosaur

In 1980, Perle and colleague Rinchen Barsbold described a second Mongolian segnosaur. In addition to vertebrae and limb bones, the skeleton includes an excellent skull, which is long-snouted with huge nostril openings, inflated bones at the base of the braincase, and a toothless beak at the front of the jaws. Barsbold and Perle named the species *Erlikosaurus andrewsi*. It is smaller than *Segnosaurus* and has more strongly compressed foot claws and a less down-curved jaw tip.

Placing the group

How these dinosaurs might be related to other forms was a problem, and several different, disparate ideas were proposed during the 1980s. In 1984, Greg Paul argued that segnosaurs resembled both prosauropods and ornithischians, and that they might be late-surviving relics of the prosauropod-

ornithischian transition (Paul thought that Saurischia was not a natural group, and that sauropodomorphs and ornithischians should be united in a group called Phytodinosauria). Other workers proposed that segnosaurs might be sauropodomorphs, but they were generally still thought of as mysterious theropods.

New species described since 1993 have shown that segnosaurs are in fact unusual, broad-footed coelurosaurs related to oviraptorosaurs. About 13 species are currently recognized. The gigantic *Therizinosaurus* turned out to be a member of the group, and because Maleev had published the name Therizinosauridae before Barsbold proposed the name segnosaur, the unwieldy name therizinosauroid was proposed for these dinosaurs in 1993.

FOLIAGE EATERS

How might segnosaurs have lived? Their broad, functionally four-toed feet suggested to some that they may have been amphibious dinosaurs of swampy environments, grabbing fish with their toothless jaw tips. The rumored presence of four-toed fossil footprints that preserved evidence of webbing provided support for this imaginative idea. In fact, the small, leaf-shaped teeth and bulky bodies of segnosaurs do not look at all suited for amphibious life. The broad belly, stout hind limbs, and leaf-shaped teeth suggest that they were herbivores that perhaps used their long arms and large hand claws to pull at foliage.

Prosauropod nests and babies

Above Like most baby animals, juveniles of *Mussaurus* had bigger heads and shorter necks than adults. They were probably quadrupedal, but this might not have been true of adults.

Prosauropods have been known to science since the 1830s, and their dinosaur group was among the first to be identified: *Plateosaurus* was one of the first dinosaurs to be given a scientific name. However, little was known of prosauropod biology, and as recently as the 1970s their nests, eggs, and babies had never been reported. So the discovery, in 1977, of what appeared to be a prosauropod nest, preserved with both eggs and recently hatched babies, was of huge importance. The specimens were discovered in the Upper Triassic Laguna Colorada Formation on an expedition of the Universidad Nacional de Tucumán and the Fundación Miguel Lillo.

Above This juvenile *Mussaurus* skeleton is still preserved in the sediment that enclosed it. Articulated baby dinosaurs skeletons are extremely rare.

Eight baby prosauropod skeletons were discovered, some of which were complete and fully articulated, and they were found in close association with two complete eggs and various eggshell fragments. These fossils belonged to a new species and, given the tiny size of the skeletons, Argentine paleontologist José Bonaparte and colleague Martin Vince decided in 1979 to name it *Mussaurus patagonicus*, which means "Patagonian mouse lizard." Ultimately, this name is misleading, because an adult *Mussaurus* was probably similar in size to other prosauropods, and the babies of other prosauropod species were probably as small as those of *Mussaurus*, as were those of many other types of dinosaur.

As expected for babies, the skulls of the *Mussaurus* juveniles have short, tall snouts and proportionally enormous eye sockets that are nearly half as long as the whole skull. However, Bonaparte and Vince thought that *Mussaurus* was so different from other prosauropods that it deserved its own group: the Mussauridae. Partly this was because they observed a peculiar and unique combination of both primitive and advanced features in the young dinosaurs.

Curious features

Bonaparte and Vince noted that the short and tall neck vertebrae resemble those of "proto-dinosaurs" such as *Lagosuchus* (now known as *Marasuchus*) more than those of prosauropods (in which the vertebrae are normally long and low); and the relatively deep, robust chin region is more characteristic of Jurassic sauropods than of Triassic prosauropods. They thought it unlikely that the short, tall skull of *Mussaurus* changed during growth into the much longer, lower skull characteristics of prosauropods such as *Anchisaurus* and *Plateosaurus*, and they thought that the teeth of *Mussaurus* are longer and more cylindrical than is typical for prosauropods. *Mussaurus*, they argued, was something quite new. However, the obvious fact that the specimens were babies with a lot of growing to do also raised the possibility that these "unique" features might change substantially during growth.

The long, cylindrical teeth, robust chin region, and short, tall skull of *Mussaurus* are all features reminiscent of sauropod skulls. Bonaparte and Vince made the intriguing proposal that sauropods might have been neotenous: that is, that they retained the juvenile features of their ancestors, the prosauropods. Other discoveries have since been made that may support this hypothesis. Hatchling specimens

of the African sauropodomorph *Massospondylus*, described in 2005, show that these babies were more like adult sauropods than adult *Massospondylus*, in terms of limb proportions.

Growth changes

A more complete analysis of *Mussaurus* was published by Argentine paleontologists Diego Pol and Jaime Powell in 2007, and by this time new remains that appear to be from juveniles or subadults were known. The new specimens have skulls about 4 in (10 cm) long, which show that as the animal grew, the snout became much longer, and the back of the skull became longer and taller. These changes are all typical of vertebrates.

Although the hatchlings are blunt-nosed, the subadults have a pointed snout tip. Possible adult specimens are also known: a large, fragmentary skull discovered in the same region is thought to be from an adult *Mussaurus*. Unfortunately, its inclusion within the species *Mussaurus patagonicus* cannot yet be confirmed because few bones are shared by the adult and by the hatchling and subadult *Mussaurus* specimens.

DISCOVERY PROFILE

Name	*Mussaurus patagonicus*
Discovered	La Colorado Lake, Santa Cruz Province, Argentina, by Martin Vince, 1977
Described	By José Bonaparte and Martin Vince, 1979
Importance	First prosauropod hatchlings and eggs to be discovered
Classification	Saurischia, Sauropodomorpha, Plateosauria

Above This is the best-known of the several baby *Mussaurus* skeletons that have been discovered. Its neck is curved backward in the death posture typical for dinosaurs.

Imaging dinosaurs

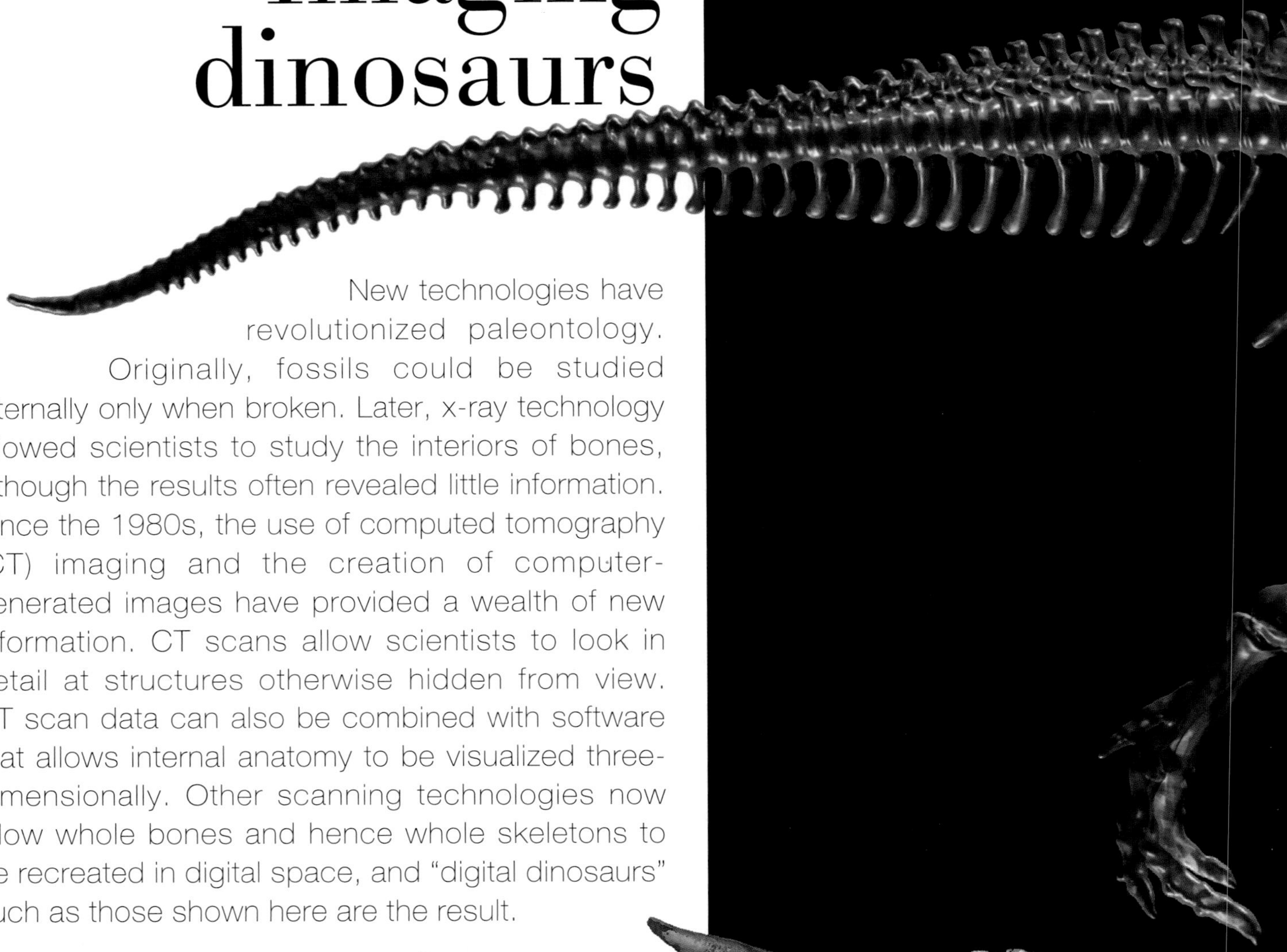

New technologies have revolutionized paleontology. Originally, fossils could be studied internally only when broken. Later, x-ray technology allowed scientists to study the interiors of bones, although the results often revealed little information. Since the 1980s, the use of computed tomography (CT) imaging and the creation of computer-generated images have provided a wealth of new information. CT scans allow scientists to look in detail at structures otherwise hidden from view. CT scan data can also be combined with software that allows internal anatomy to be visualized three-dimensionally. Other scanning technologies now allow whole bones and hence whole skeletons to be recreated in digital space, and "digital dinosaurs" such as those shown here are the result.

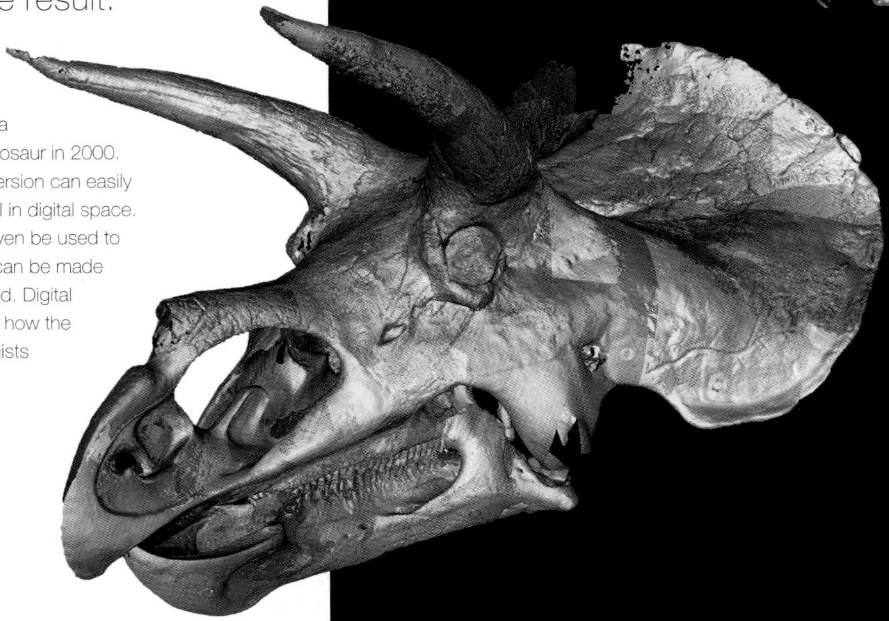

Right By scanning the skull and skeleton of a *Triceratops*, a museum team succeeded in making the very first digital dinosaur in 2000. Unlike the real skull, which is huge, bulky, and fragile, this version can easily be flipped, rotated, manipulated, and examined internally, all in digital space. Using a technique called prototyping, the digital data can even be used to create physical copies of the scanned fossil. Such copies can be made of plastic or paper and produced at whatever size is required. Digital models can also be subjected to simulated stresses to see how the real objects might have performed in life. Some paleontologists regularly use these digital models to analyze such areas as feeding behavior and locomotion.

3D DIGITAL IMAGE OF *TYRANNOSAURUS* SINUSES

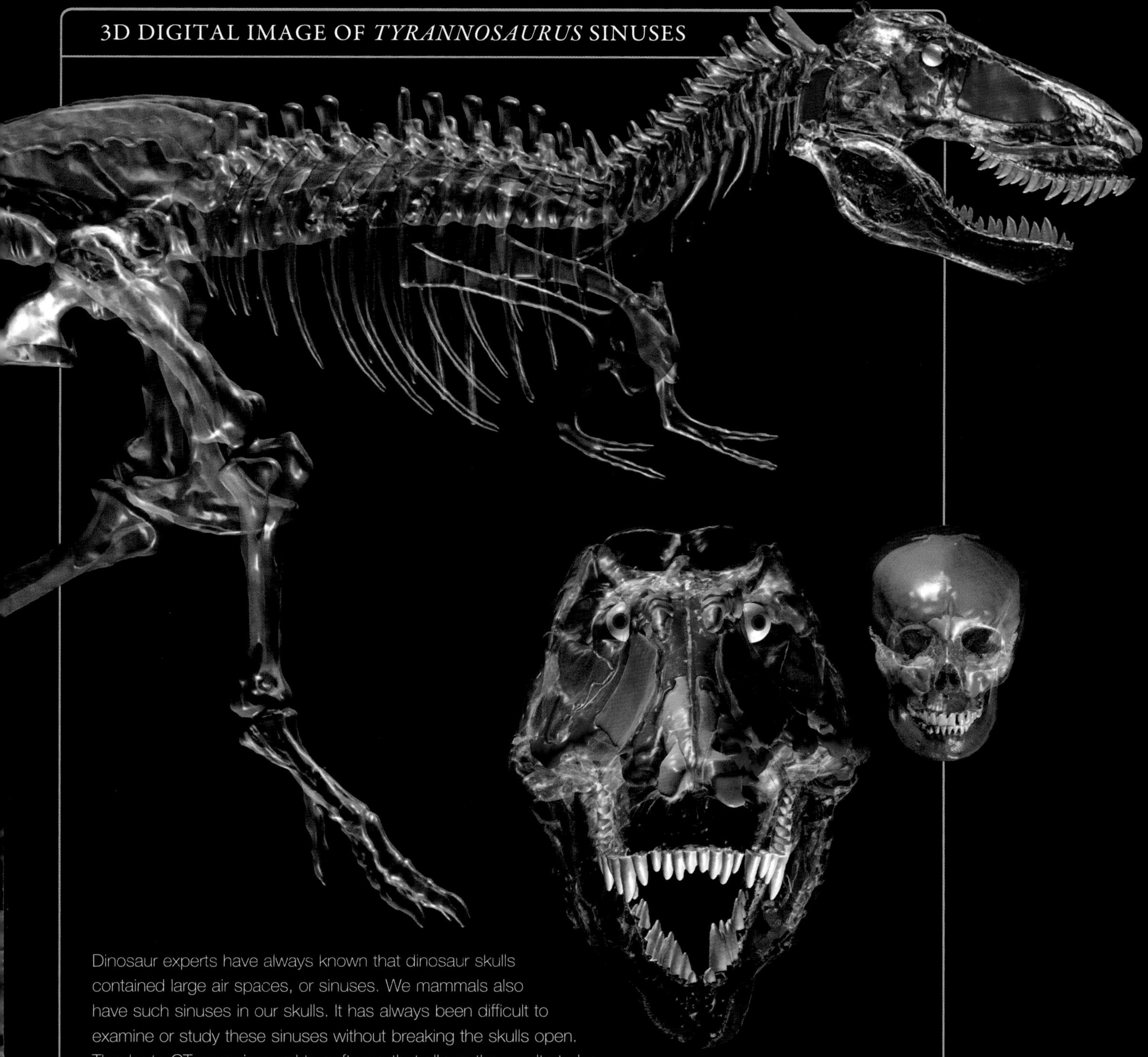

Dinosaur experts have always known that dinosaur skulls contained large air spaces, or sinuses. We mammals also have such sinuses in our skulls. It has always been difficult to examine or study these sinuses without breaking the skulls open. Thanks to CT scanning and to software that allows the results to be visualized, experts have recently been able to image dinosaur sinuses in incredible detail. Large theropods such as *Tyrannosaurus*, shown here, are "airheads," with about 18 percent of the skull's mass being occupied by air. Air-filled cavities occupied most of the snout and palate bones, and as many as ten distinct sinuses branched off from these to occupy other regions of the skull. Now that this complex system has been mapped in detail, a large question remains: what were these air spaces for? Perhaps the sinuses were advantageous because they made the skulls lighter or stronger. However, it is also possible that they played a role in temperature regulation, or even in vocalization.

"Heavy claw" and the spinosaurs

England is the scientific "home" of the dinosaurs and continues to yield new species. One of the most surprising is a new large theropod discovered in Lower Cretaceous Wealden sediments at a clay pit in Ockley, Surrey, in 1983, by amateur collector William Walker. He discovered parts of a giant claw, a claw bone, and a tail bone.

Above The gigantic thumb claw of *Baryonyx*, here reconstructed with its horny sheath, was strongly curved and sharply pointed.

After repairing the broken tip of the large claw, Walker realized that these fossils belonged to something quite unusual. He took them to the British Museum (Natural History) in London, which today is called the Natural History Museum. Here they were examined by fossil reptile experts Alan Charig and Angela Milner. This was clearly an important discovery, and a visit to the clay pit revealed further bones that seemed to belong to the same animal.

A full excavation

The resulting excavation and preparation revealed much of the skull and a partial skeleton from an animal perhaps 30 ft (9 m) long. The huge claw, which measures over 12 in (30 cm) along its curve, belonged to the hand, and the skull is narrow and crocodile-like, with an expanded snout tip. The animal appeared to be a highly unusual theropod. However, some paleontologists suggested that it might be a crocodilian or even a late-surviving rauisuchian (a group of non-dinosaurian archosaurs otherwise restricted to the Triassic).

In 1986, Charig and Milner published a preliminary description. The new dinosaur was definitely a theropod after all, and they named it *Baryonyx walkeri*, meaning "Walker's heavy claw." It was regarded as so unusual that it was worthy of its own family, which they named Baryonychidae. The rocks of the Wealden have been studied and collected from for over 150 years, so the discovery of a completely new kind of large predatory dinosaur in the English Wealden was nothing short of amazing. The media soon nicknamed *Baryonyx* "Claws" in a punning reference to the popular film *Jaws*, and the dinosaur became an international sensation.

Left Few amateur collectors get to make discoveries as exciting as that made by William Walker, shown here holding his well-preserved and amazing find.

DISCOVERY PROFILE

Name	*Baryonyx walkeri*
Discovered	Ockley, Surrey, England, by William Walker, 1983
Described	By Alan Charig and Angela Milner, 1986
Importance	The first good spinosaurid skeleton, and the first spinosaurid from Europe
Classification	Saurischia, Theropoda, Spinosauroidea, Spinosauridae

Massive arms

Baryonyx had a massively muscled upper arm bone, and this caused Charig and Milner to suggest that it crouched or walked quadrupedally, making the species unique among known theropods. Its crocodile-like skull, large thumb claws, and finely serrated teeth indicate that it was a fish eater, and this was confirmed by the discovery of fossil fish scales preserved in its stomach region.

An alternative hypothesis, proposed in 1987, was that *Baryonyx* might be a specialized scavenger that used its huge claw to rip carcasses open and its long snout to probe into them. Although this behavior is plausible, the idea that a giant terrestrial carnivore could find enough carcasses to survive is unlikely.

Solving the enigma

The affinities of *Baryonyx* proved controversial. Charig and Milner stated that too little was known to come to any conclusions, but they did note that details of the snout bones vaguely recalled those of *Dilophosaurus* and similar Early Jurassic theropods. However, during the late 1980s, French paleontologist Eric Buffetaut and American writer and illustrator Greg Paul noted that the remains of *Baryonyx* seemed to be quite similar to those of the enigmatic African theropod *Spinosaurus aegyptiacus*. Both dinosaurs have an unusual upturned tip to the lower jaw, and also share features of the teeth and tooth sockets. Accordingly, *Baryonyx* was not such an enigma after all, but a member of Spinosauridae.

New fossils and analysis have shown that this theory is correct. A close relative of *Baryonyx*, *Suchomimus*, was named from Niger in 1998, a Brazilian spinosaurid (*Irritator challengeri*) was named in 1994, and numerous fragmentary fossils of *Baryonyx* (mostly teeth) have been reported from England, Germany, Spain, and Portugal. All spinosaurids seem to have been long-jawed predators that behaved quite differently from deep-skulled theropods, such as megalosaurs and allosaurs. Rather than providing evidence for quadrupedality, the robust arm bones of these theropods may have helped them to tackle or pick up prey.

Above *Baryonyx* and related spinosaurids were large theropods with crocodile-like skulls and powerful arms. They are thought to have eaten a lot of fish.

117

Jim Jensen's giants

During the 19th century, the productive Upper Jurassic rocks of the Morrison Formation had yielded *Camarasaurus*, *Apatosaurus*, *Diplodocus*, and *Barosaurus*, and then, in 1903, came the gigantic *Brachiosaurus*. For most of the 20th century, it must have seemed that the Morrison Formation had no new major sauropods to reveal. Nevertheless, in 1972 remains from a new sauropod were found in the Morrison Formation, and the discovery was made all the more exciting by the fact that this animal was reported to exceed all other sauropods in size.

These fossils were reported by Jim Jensen, of Utah's Brigham Young University, who collected them from Dry Mesa quarry in Colorado in 1979. The sauropods were found at the same locality as a geological feature called the Uncompahgre Upwarp, and they were discovered alongside a rich fauna of small dinosaurs and pterosaurs, on which Jensen published throughout his career.

Jensen identified a gigantic scapulocoracoid (the unit of the pectoral girdle that contains the coracoid and scapula) found at Dry Mesa as a new sauropod of unknown affinities. The scapulocoracoid was 8 ft (2.4 m) long. In magazine articles, he took to referring to the sauropod as

Left Jensen poses here with a reconstructed forelimb of *Ultrasauros*. Because *Ultrasauros* was thought to be a giant relative of *Brachiosaurus*, the reconstruction is essentially a scaled-up version of that dinosaur's arm.

"*Supersaurus*," and in 1985 he formally published the name as *Supersaurus vivianae*. The specific name honored Vivian Jones, who, together with Eddie Jones, had discovered the Uncompahgre sites when prospecting for uranium in 1943. *Supersaurus* appeared to be similar to *Diplodocus* and *Apatosaurus*. A second, larger scapulocoracoid almost 9 ft (2.7 m) long, together with some pelvic bones and vertebrae, were thought to belong to it as well. *Supersaurus* is a diplodocid, and hence is long-necked and long-tailed, even for a sauropod. In fact, the animal's total length was estimated to be up to 147 ft (45 m).

More massive bones

Jensen thought that the remains of another enormous sauropod were present in the assemblage. Lying in the middle of the *Supersaurus* bones was a gigantic dorsal vertebra a little over 4 ft (1.3 m) tall that, he argued, seemed to belong to a brachiosaur.

DISCOVERY PROFILE

Name	*Supersaurus vivianae*
Discovered	Dry Mesa quarry, Mesa County, Colorado, U.S.A., by James Jensen, *c.*1972
Described	By James Jensen, 1985
Importance	Best known gigantic North American sauropod
Classification	Saurischia, Sauropodomorpha, Sauropoda, Diplodocoidea

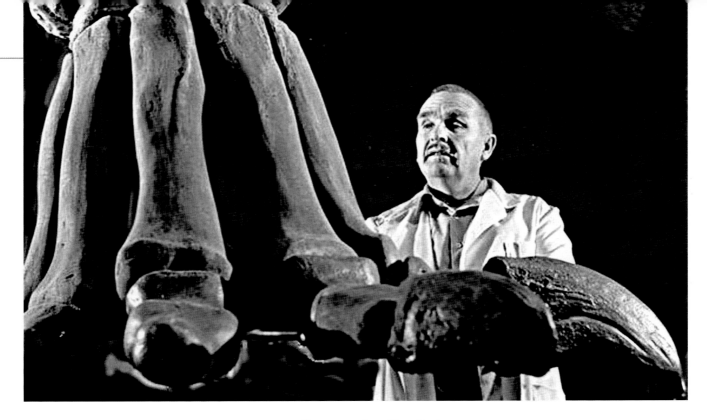

Jensen unofficially termed it "*Ultrasaurus*" and it was mentioned in numerous magazine and newspaper articles before its official publication as *Ultrasaurus macintoshi* in 1985 (later changed to *Ultrasauros*). Jensen assumed that yet another vast scapulocoracoid found nearby, along with one neck and one tail vertebra, also belonged to *Ultrasauros*.

Jensen's third Uncompahgre sauropod, *Dystylosaurus edwini*, was based on a single dorsal vertebra that was thought to be so distinctive that, wrote Jensen, it "no doubt represents a new sauropod family." Perhaps because of its more awkward name, *Dystylosaurus* never won the popularity awarded to *Supersaurus* and *Ultrasauros*. During the 1970s and 1980s, the latter two species appeared in almost every popular dinosaur book and article, where they were generally proclaimed to represent the longest and heaviest dinosaurs of them all. *Ultrasauros* was imagined as a scaled-up *Brachiosaurus*, and was even given an estimated weight of up to 200 tons (180 tonnes).

Reassessing the finds

Jensen's sauropods were re-evaluated by Brian Curtice and colleagues during the 1990s and early 21st century. The *Ultrasauros* vertebra is not from a brachiosaur after all, but from a diplodocid, and almost certainly just another bone from *Supersaurus*. The giant *Ultrasauros* scapulocoracoid is, however, from a brachiosaur. *Dystylosaurus* also proved less unusual than Jensen thought, and in 2001 it was also argued to belong to *Supersaurus*.

A new specimen of *Supersaurus* was described in 2007. Analysis of its characters has shown that it shares features with *Apatosaurus*, and, accordingly, it is regarded as a member of the diplodocid clade Apatosaurinae. To date, apatosaurines are exclusively North American, and it has been suggested that they were endemic to the continent. In contrast to *Apatosaurus* (the best known apatosaurine), *Supersaurus* has slender, narrow neck vertebrae. In fact, its neck is the longest yet reported for any sauropod, with an estimated length of over 43 ft (13 m).

Above The American paleontologist "Dinosaur" Jim Jensen (1919–1998) was best known for his work on gigantic Jurassic sauropods, but he discovered many other exciting Jurassic fossils as well as specimens from the Triassic and Cenozoic.

Left With an estimated length of 115–150 ft (35–45 m), *Supersaurus* is one of the longest sauropods. It is closely related to *Apatosaurus*.

119

The *Chasmosaurus* bonebed

The Big Bend region in Texas has yielded dinosaur remains since the early 20th century, but many remained unexamined. In 1989, paleontologist Thomas Lehman named a new ceratopsid ceratopsian known from more than 300 stored bones belonging to many individuals. He named it *Chasmosaurus mariscalensis*, including it in the same genus as *C. russelli* and *C. belli* from Canada.

Above The unusually erect frill of the Big Bend ceratopsid *Chasmosaurus mariscalensis* is obvious in this artistic reconstruction. Like all ceratopsids, it was equipped with a massive beak and shearing cheek teeth.

Below The massive window-like openings that lightened the frills of most ceratopsids are particularly well displayed in this skeleton of the Canadian species *Chasmosaurus belli*.

Compared to other *Chasmosaurus* species, *C. mariscalensis* has longer brow horns and a broader squamosal (the bone that makes up the side part of the frill), and it also seemed to have only six epoccipitals along the frill's side border (epoccipitals are triangular bones arranged around the edges of the frill, and in other *Chasmosaurus* species there are between seven and nine of them on each frill edge).

In 1993, a complete Big Bend *C. mariscalensis* skull was described by Catherine Forster and colleagues. Journalists claimed that this 5 ft (1.5 m) long specimen was "the largest skull of any land animal ever found," but they were mistaken. Other large ceratopsids with longer skulls had long been known, such as *Torosaurus*, which has a skull is up to 8½ ft (2.6 m) long. The Big Bend skull showed how *C. mariscalensis* has a strongly compressed nose horn, and also that Lehman's ideas about a low epoccipital number were incorrect: large individuals of *C. mariscalensis* in fact have as many as 10 of these bones along each side of the frill.

Long horns

An analysis of the new skull enabled Forster and colleagues to argue that the long horns of *C. mariscalensis* might not be an unusual feature, as originally thought. Instead, it seemed that short horns (as seen in *C. russelli* and *C. belli*) were unusual, and that long horns were the primitive, "normal" condition for this group of ceratopsids. Indeed, the idea that long horns are primitive for *Chasmosaurus* is supported by the fact that *Chasmosaurus* seems most closely related to *Pentaceratops* (which is long-horned), and that other chasmosaurines (such as *Triceratops*, *Torosaurus*, and *Anchiceratops*) are also long-horned.

DISCOVERY PROFILE

Name	*Chasmosaurus mariscalensis*
Discovered	Big Bend, Texas, U.S.A., by William Strain and colleagues, 1938–39
Described	Thomas Lehman, 1989
Importance	One of best represented ceratopsids; revealed key information on variation within a species
Classification	Ornithischia, Marginocephalia, Ceratopsia, Ceratopsidae

An old specimen

C. mariscalensis was not actually a new discovery. It came from the Upper Cretaceous Aguja Formation and was excavated in 1938–39 as part of a Works Progress Administration project supervised by William Strain of the College of Mines and Metallurgy Museum at El Paso. Strain had produced a report mentioning the dinosaurs, but nothing else was done with them for decades and they remained in storage. At about the same time, an expedition from the University of Oklahoma set out to collect fossils from Big Bend, and in 1938 Wann Langston and colleagues recovered the fragmentary remains of a large ceratopsid. This specimen proved particularly important, but it also was placed in storage and published data did not appear on it until 2007, when it was recognized as one of the most complete members of its species.

Sorting the species

Chasmosaurus had originally been named by Lawrence Lambe in 1914 for *C. belli* from Alberta, a species he had first named in 1902 as *Monoclonius belli*. All *Chasmosaurus* species have a very long frill that is essentially rectangular in shape, perforated by enormous window-like openings, and decorated around its edges by epoccipitals. Some species have short horns over the eyes and others have long ones, and a species from Alberta named in 2001, *C. irvinensis*, lacks brow horns altogether. *C. mariscalensis* differs in details of its snout anatomy from other *Chasmosaurus* species and one study concluded that it might actually be more closely related to *Pentaceratops*. If this is correct, then it deserves to be separated from *Chasmosaurus* and given its own genus. In 2006, one group of researchers did precisely this, and proposed the name *Agujaceratops* for the species.

Left The long, erect frill of *Chasmosaurus mariscalensis* could well have been used in display. If so, bold patterns or colors may have decorated its surface.

VARIATION WITHIN SPECIES

In 1990, Lehman published a study on the degree of variation within *Chasmosaurus mariscalensis*. Among the 10 preserved individuals, the horns differed in size and shape, as did the length of the face. Did this indicate that more than one species was preserved, or that individuals within a ceratopsid species could be quite variable? Studies of growth and variation within horned dinosaurs indicate that these details were variable within species. However, Lehman's argument that male and female ceratopsids can be distinguished on the basis of horn and skull shape has proved controversial, and the supposed sexual differences that have been described most likely represent the sort of variation normal within a population.

Feathers and Fur: A New Diversity

Inspired by the discoveries of the 1960s and 70s, American, European, and Argentinean research teams carried out fieldwork in Mongolia, Africa, and Argentina during the 1990s. South America proved particularly important, and a haul of remarkable new dinosaurs shed much new light on poorly known regions of the dinosaur family tree. The number of recognized dinosaurs has undergone an extraordinary 85 percent increase since 1990. The study of dinosaurs had clearly entered a new "golden age."

Facing page During the 1990s, new fieldwork programs in Africa, Mongolia, China, and elsewhere resulted in the discovery of dozens of new dinosaurs. This striking Mongolian dromaeosaur was discovered on a 1993 expedition. Originally assumed to be a *Velociraptor*, it was shown in 2006 to represent a new species: *Tsaagan mangas*.

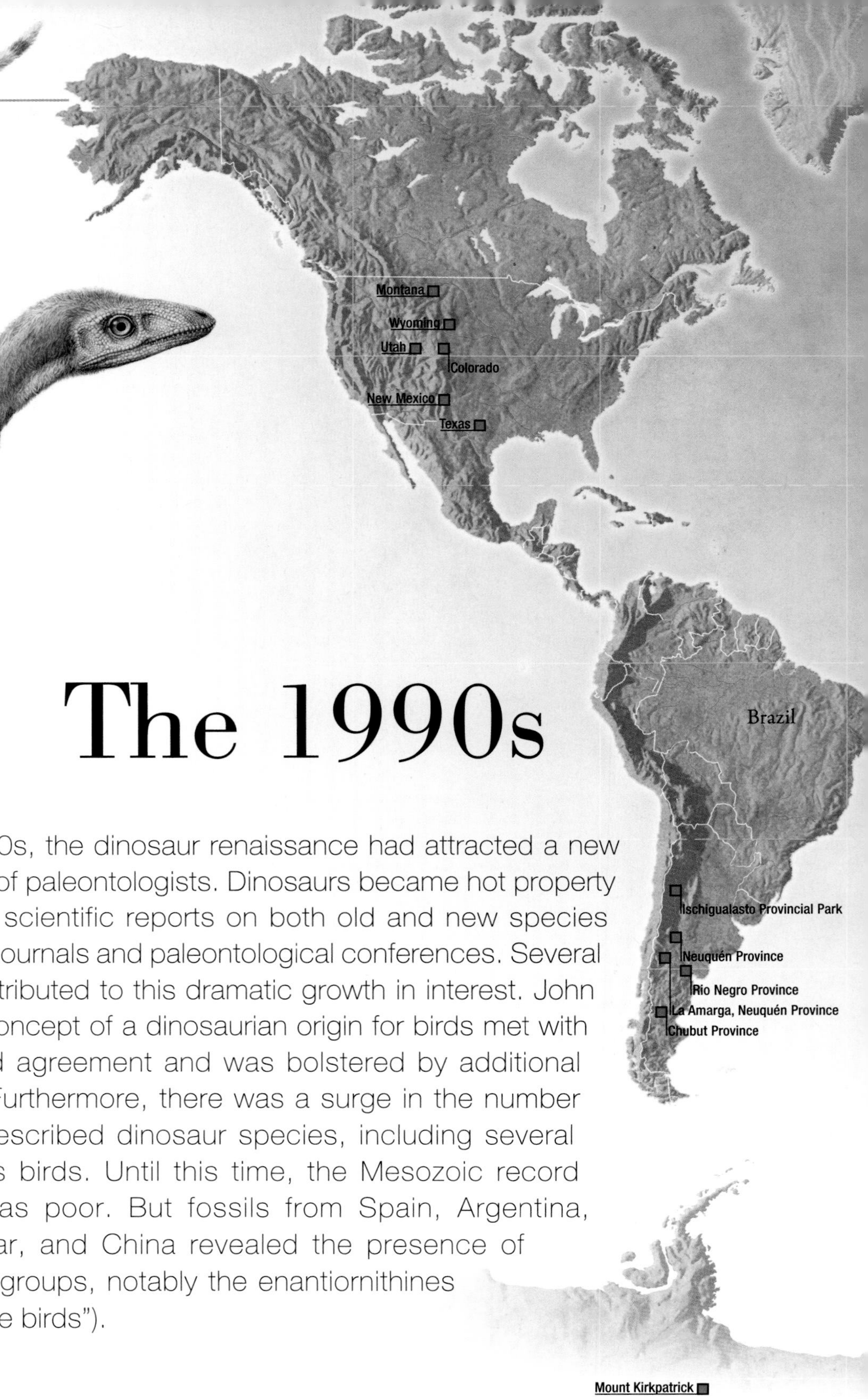

The 1990s

Above The idea that some dinosaurs were feathered is very old, but during the 1990s these speculations were confirmed. The first of these feathered dinosaurs was little *Sinosauropteryx*, which came from China. Quill-like "proto-feathers" covered its body. These are almost certainly the structural ancestors of the true feathers that evolved later on.

By the 1990s, the dinosaur renaissance had attracted a new generation of paleontologists. Dinosaurs became hot property again, and scientific reports on both old and new species dominated journals and paleontological conferences. Several factors contributed to this dramatic growth in interest. John Ostrom's concept of a dinosaurian origin for birds met with widespread agreement and was bolstered by additional research. Furthermore, there was a surge in the number of newly described dinosaur species, including several Cretaceous birds. Until this time, the Mesozoic record of birds was poor. But fossils from Spain, Argentina, Madagascar, and China revealed the presence of entire new groups, notably the enantiornithines (or "opposite birds").

Montana
Wyoming
Utah
Colorado
New Mexico
Texas

Brazil

Ischigualasto Provincial Park
Neuquén Province
Rio Negro Province
La Amarga, Neuquén Province
Chubut Province

Mount Kirkpatrick

1990 *Carnotaurus* described in full detail

1991 Spiky sauropod *Amargasaurus* is named

1993 Early saurischian *Eoraptor* is named
1993 *Mononykus* (originally named *Mononychus*) is named
1993 Gigantic sauropod *Argentinosaurus* is named
1993 *Utahraptor* is published

1992 Horner and colleagues refer to new ceratopsids from Montana

1994 *Afrovenator* described from Niger
1994 *Pelecanimimus* named from Spain
1994 A strange theropod named: *Cryolophosaurus*

1995 The new ceratopsids from Montana are named: *Einiosaurus* and *Achelousaurus*
1995 Patagonian giant theropod *Giganotosaurus* is named

1990   1991   1992   1993   1994   1995

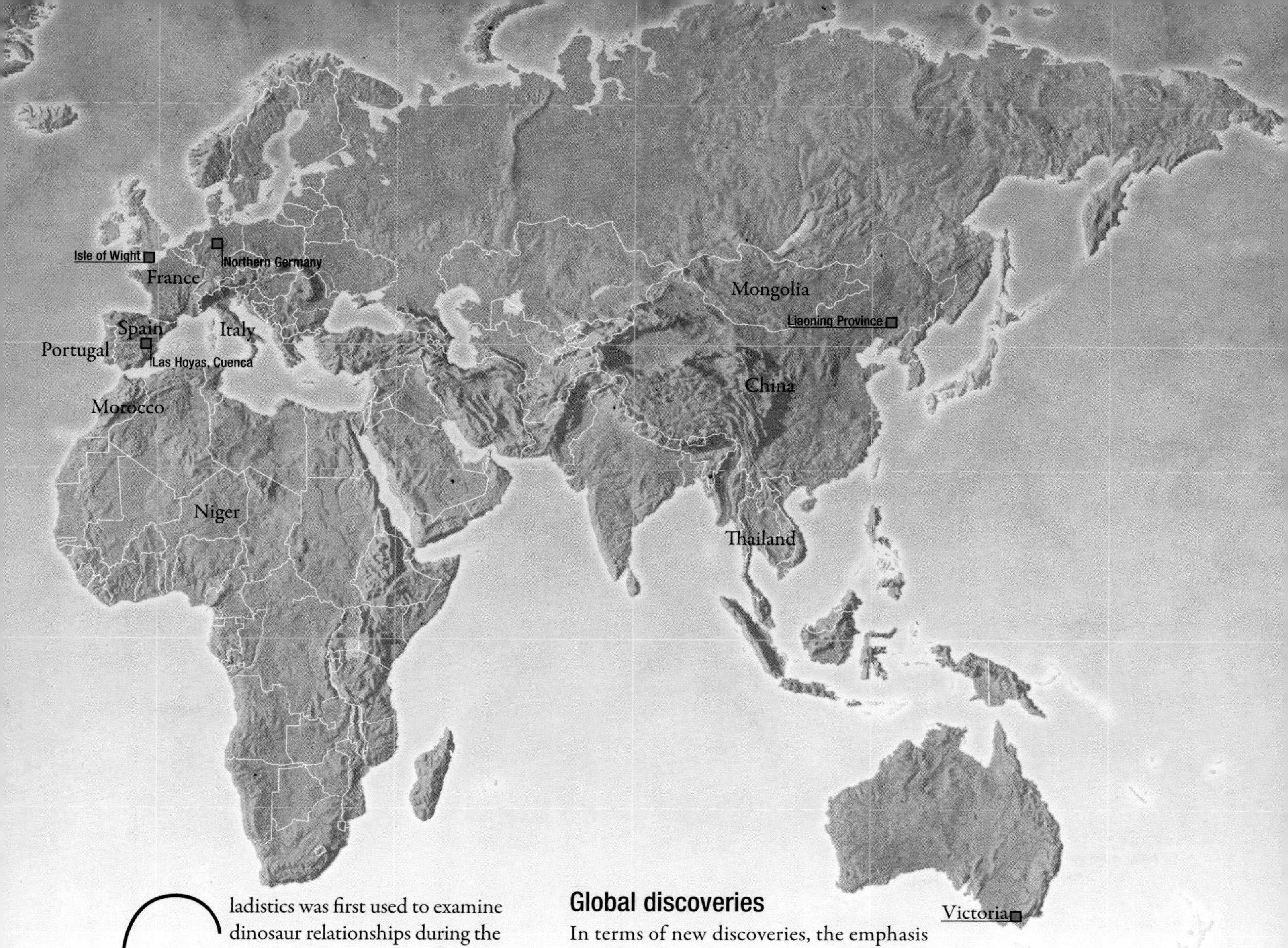

Isle of Wight

Northern Germany

France

Mongolia

Spain   Italy

Portugal

Las Hoyas, Cuenca

Liaoning Province

Morocco

China

Niger

Thailand

Victoria

Global discoveries

Cladistics was first used to examine dinosaur relationships during the mid-1980s. It was by the 1990s the standard technique for analyzing dinosaur phylogeny, and a well-resolved family tree had been developed for the whole of the Dinosauria. There was much interest in testing Ostrom's and Bakker's ideas about warm-bloodedness in dinosaurs further, but it was argued that many of the claims made by Bakker could not definitely be linked to physiology and thus were inconclusive. New work on the internal anatomy of dinosaur bone by Robin Reid, Armand de Ricqlès, and others showed that dinosaurs grew quickly. Ultimately, however, the "warm-blooded" debate continued.

In terms of new discoveries, the emphasis shifted as dinosaurs from South America, Africa, Madagascar, and China became increasingly important in shaping views about evolution and diversity. This is not to say that the better known regions, North America and Europe, stopped producing new data and new species. Working mostly in the Cretaceous rocks of Utah, United States, Jim Kirkland and colleagues discovered dinosaurs that were either entirely new, or represented close relatives of types that were already known from contemporary rocks in Europe. The giant, sickle-clawed dromaeosaur *Utahraptor* was named in 1993, for example.

1996 Another alvarezsaurid, *Patagonykus*, is described
1996 *Neovenator* from England is finally named
1996 Sereno and colleagues report new *Carcharodontosaurus* material
1996 The feathered compsognathid *Sinosauropteryx* is announced

1999 *Nigersaurus* and *Jobaria* are named

1998 *Nedcolbertia*, *Eolambia*, and *Gastonia* from Utah are published
1998 Feathered theropods *Caudipteryx* and *Protarchaeopteryx* described
1998 Important early horned ceratopsian: *Zuniceratops*

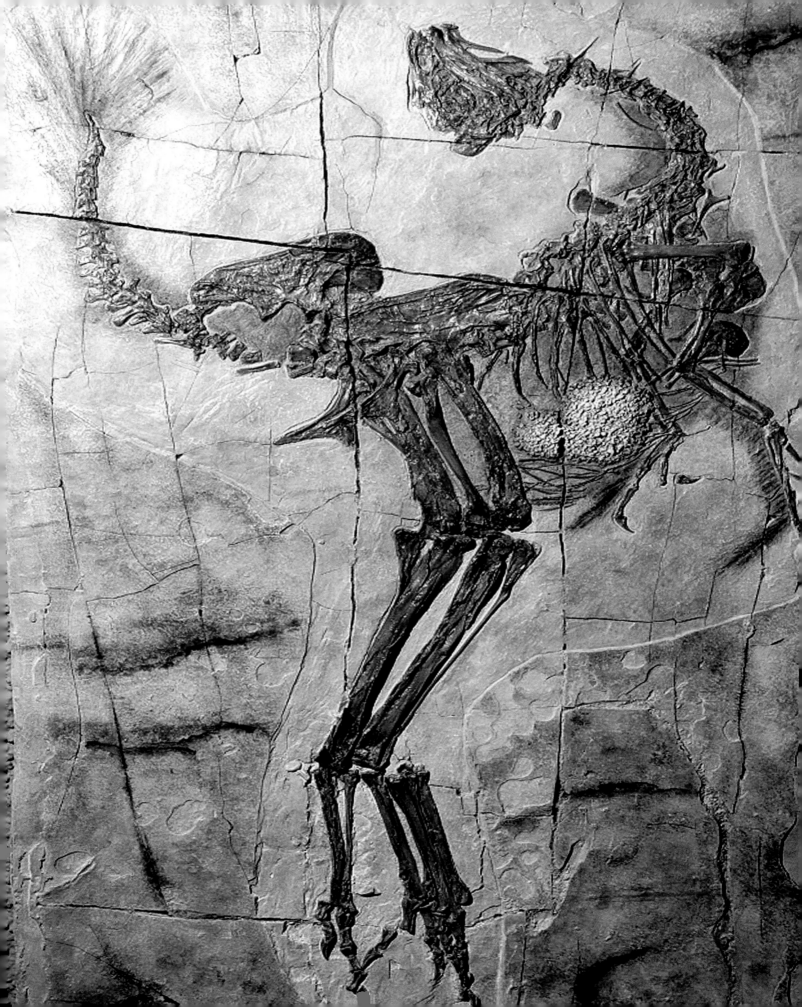

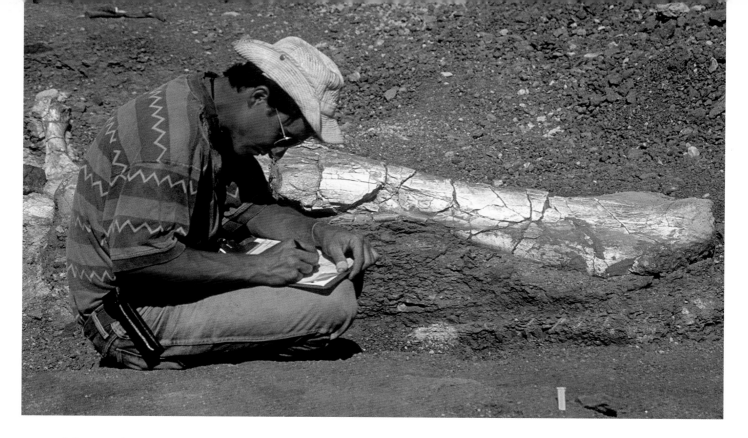

Other North American discoveries included the small theropod *Nedcolbertia*, ankylosaur *Gastonia*, iguanodontian *Eolambia*, and new ceratopsian *Zuniceratops*, which were all published in 1998. Several new North American sauropods recognized during this time indicated that the continent's Cretaceous sauropod record was richer than people realized.

South America produced a slew of new dinosaurs. Mostly this was due to the research efforts of a few individuals, particularly José Bonaparte and his students and colleagues. The list of new South American sauropods named during the 1990s is impressive, and includes *Epachthosaurus*, *Amargasaurus*, *Andesaurus*, *Neuquensaurus*, *Argentinosaurus*, *Rayososaurus*, *Pellegrinosaurus*, *Agustinia*, *Gondwanatitan*, and *Tehuelchesaurus*. The abelisaurid theropods, first recognized during the 1980s, became better known. Horned *Carnotaurus* was described in full in 1990 and it seemed that these dinosaurs were present in Europe and India as well as South America. *Majungasaurus* from Madagascar was also an abelisaurid, and it proved to be the best preserved by far.

Fruitful expeditions

After 1990, the American Museum of Natural History once again resumed fieldwork in Mongolia, and some of the results were spectacular. New oviraptorosaurs, troodontids, and dromaeosaurs were discovered. The famous "*Protoceratops*" eggs found in the 1920s were re-identified as oviraptorosaur eggs. Paul Sereno from the University of Chicago led expeditions to Niger and Morocco: new dinosaurs were discovered, but so was new material of animals originally described by Ernst Stromer back in the early 20th century. By far the most newsworthy dinosaurs of the time, however, came from Liaoning Province in northeast China. In 1996, a small Liaoning theropod called *Sinosauropteryx* was reported. It was covered in simple, hair-like structures, apparently the structural antecedents to true feathers. In 1998, this fascinating species was followed by the indisputably feathered *Caudipteryx* and *Protarchaeopteryx*.

During the 1960s and 70s, the ideas of the dinosaur renaissance were rapidly absorbed into popular culture, thanks to a proliferation of books, magazine articles, and artwork. Dinosaurs increasingly entered mainstream culture throughout the 1980s and 90s, and 1993 saw the screening of *Jurassic Park*, based on Michael Crichton's 1990 novel. It would be wrong to imply that either the book or film influenced scientific thought, but there is no denying the fact that *Jurassic Park* brought modern views about dinosaurs to an eager new audience.

Above During the 1990s, major efforts by American paleontologist Paul Sereno and colleagues uncovered new dinosaurs from Africa, the "dark continent" for dinosaur research. Here, a member of Sereno's team takes notes at a field site.

Facing page *Caudipteryx*, named in 1998, was a turkey-sized relative of *Oviraptor*. It was among the first of many spectacular new dinosaurs that emerged from Liaoning Province in China.

An exquisite
Australian ankylosaur

Most ankylosaur fossils are from the Northern Hemisphere. A few specimens are known from the southern continents that once formed the supercontinent Gondwana, but these are mostly fragmentary. By far the best known ankylosaur from Gondwana is little *Minmi paravertebra* from Australia. It was named in 1980 by Ralph Molnar and was initially referred to as "the Minmi ankylosaur."

Below Spikes and armor plates projected from the hind limbs, tail, and body of *Minmi*. The tail spikes projected sideways and made the tail a formidable weapon.

The new Australian dinosaur was named for Minmi Crossing in Queensland. Its species name, *paravertebra*, refers to bones called paravertebrae arranged along the sides of the tops of its vertebrae. Molnar made the intriguing suggestion that the paravertebrae may have been attached to the armor on the top of *Minmi*'s back to allow the armor to be raised in a threat or defensive display. This idea could not be supported, however.

In a 1987 article, Molnar and German crocodilian expert Eberhard Frey proposed that the paravertebrae played a crucial role in supporting the musculature of the back. They suggested that *Minmi* had a reduced complement of armor, and that a strong back musculature showed that it relied on speed to escape from predators. If this is correct, then *Minmi* was, so far as we know, a truly unusual ankylosaur.

The original *Minmi* specimen consists of vertebrae, ribs, a hand, and armor. A second, more complete skeleton was discovered in 1989

DIET INSIGHT

The 1989 *Minmi* specimen was of particular interest to researchers, because it preserved gut contents. These were described fully in 2000. They revealed that the animal had been eating soft leaves and fruit-like structures.

DISCOVERY PROFILE

Name	*Minmi paravertebra*
Discovered	Near Minmi Crossing, Queensland, Australia, by Alan Bartholomai, 1964
Described	By Ralph Molnar, 1980
Importance	First Australian ankylosaur; best known ankylosaur from Gondwanan continents
Classification	Ornithischia, Thyreophora, Ankylosauria

by Ian Ievers. It was geologically younger than the first specimen and could not definitely be assigned to the same species, so has therefore been classified as "*Minmi* sp."

Resting upside-down

The Ievers specimen was discovered upside-down in marine rocks, as is the case with many ankylosaur remains. Paleontologists have suggested various explanations for why ankylosaurs are often preserved this way. One proposal, advanced during the 1930s, is that ankylosaurs were marine animals, and that they fed on seaweed in shallow coastal waters. This is unlikely, because the animals lack any features consistent with this lifestyle. It is more likely that some ankylosaurs inhabited coastal plains and similar environments, and their bodies were often transported to the sea by rivers. The heavy armor and wide bodies of ankylosaurs may explain why their carcasses were often preserved upside-down.

Size issues

Minmi is about 8–11½ ft (2.4–3.5 m) in length, making it quite small for an ankylosaur. The bones of the second specimen indicate that it was adult, or near-adult, when it died. When *Minmi* lived, much of Australia was covered by shallow seas and Queensland was a set of large islands. It is possible that *Minmi* was a dwarf, island-dwelling ankylosaur much like the European *Struthiosaurus*, but this remains speculative; perhaps *Minmi* was small simply because it retained the body size that was primitive for ankylosaurs. Like all ankylosaurs, *Minmi* has a squat body that is rounded in cross-section, a short neck, and short, stocky limbs.

Articulated plates

The 1989 specimen preserves most of its armor in articulation, and this shows that the animal's upper surface, the upper parts of its limbs, and the base of its tail were covered with a pavement of small, bony plates. Larger plates were arranged on the neck and shoulders, and there were short spikes at the back of the hip region and thorn-like plates on the limbs and either side of the tail. The animal's skull is broad and deep, with proportionally large eye sockets. Ankylosaurs generally have small eye sockets and were thought to have poor vision, and although it is possible that *Minmi* was an exception to this rule, we do not know enough about its sensory abilities to reach any further conclusions.

129

The most ancient of dinosaurs

The early history of dinosaurs was always poorly known, and most ideas about dinosaur origins and diversification were based on animals such as *Coelophysis* (*see* pp.84–85). During the 1960s, however, Argentinean fossils opened up an exciting new chapter in our knowledge of primitive dinosaurs.

Above *Herrerasaurus* was one of the first dinosaurs to evolve large body size and to become a macropredator: that is, a predator of other large animals. It was probably dangerous to the much smaller *Eoraptor*.

Below This reconstructed skeleton shows that *Eoraptor* was theropod-like in overall shape and proportions. Primitive ornithischians and sauropodomorphs were this shape too.

Between 1959 and 1962, Osvaldo Reig and his colleagues found several important specimens of an early dinosaur in the Upper Triassic Ischigualasto Formation of northwest Argentina. The best specimen was discovered by Victorino Herrera in 1961, and in his honor the species was named *Herrerasaurus ischigualastensis* (the species name means "from Ischigualasto"). *Herrerasaurus* appeared to be a large, bipedal predator, but it displayed a confusing combination of features. Though Reig regarded it as a megalosaur-type theropod, he noted that it also has prosauropod-like features.

In 1970, another *Herrerasaurus*-like Triassic dinosaur was reported, this time from Brazil. American Edwin Colbert named it *Staurikosaurus* and regarded it as a relative of

Herrerasaurus. Colbert imagined these dinosaurs as bipedal, predatory saurischians, but classified them within a new group (termed Teratosauria) most closely related to prosauropods. Unlike theropods, in which the first metatarsal has become reduced and no longer makes contact with the ankle joint, *Herrerasaurus* has a primitive foot in which all five metatarsals reach the ankle.

Unusual pubis

Herrerasaurus was surprisingly specialized in parts of its anatomy. As an early saurischian dinosaur, it might be expected that its pubic bones are rod-like, with the shafts pointing forward and downward. In fact, the *Herrerasaurus* pubis is near-vertical and has a massively enlarged "boot" at its lower end. The pubis is also very broad—so much so that some

experts thought it unlikely that *Herrerasaurus* could walk with its thighs held close in to its body. In 1978, South African paleontologist Jacques Van Heerden stated that *Herrerasaurus* could not be bipedal for this reason, and he regarded *Herrerasaurus* as a primitive sauropod with splayed hind limbs. The theropod-like teeth thought by Reig to be from *Herrerasaurus* might, Van Heerden suggested, have come from another animal. These speculative proposals—made when little was known of these dinosaurs—were later falsified by new discoveries.

Predatory claws

Herrerasaurus and *Staurikosaurus* remained enigmatic, and nobody really understood how they fitted into the story of saurischian evolution. However, in 1988 the discovery of better-preserved *Herrerasaurus* remains cleared up most of the mystery. An excellent complete skull showed that *Herrerasaurus* was undoubtedly a big-toothed predator with powerful arms. The three large fingers on the inside of the hand bear huge, curved claws that could be used in grabbing or raking. The third finger was longest, and the fourth and fifth fingers were tiny and probably buried in soft tissue at the side of the palm.

The most primitive

In 1993, another primitive South American dinosaur was named: *Eoraptor lunensis*. Its describers were the American Paul Sereno

and colleagues, who argued that the animal was one of the most primitive dinosaurs yet found. *Eoraptor* was small, at just over 3 ft (1 m) long, and its limb proportions suggest that it was an agile, theropod-like biped. Like *Herrerasaurus*, it has five fingers, again with the fourth and fifth fingers reduced to tiny, clawless stumps, and the three inner fingers are long and have large, curved claws. Its skull appears rather nondescript, although its nostril opening is large and it clearly had the jaws and teeth of a predator.

Eoraptor was described by Sereno and his colleagues as a theropod, and the most primitive theropod of them all. They also suggested that *Herrerasaurus* and *Staurikosaurus*, which by this time were grouped together as the herrerasaurids, were primitive theropods too. However, since 2000 several studies on early dinosaur evolution have found both *Eoraptor* and the herrerasaurids to be primitive members of Saurischia that are outside of a clade that includes theropods and sauropodomorphs.

Above *Eoraptor* may be one of the most primitive of all dinosaurs, and its skull appears to combine features of the different dinosaur groups. The curved, blade-like teeth are theropod-like, but it also has similarities with sauropodomorphs such as *Pantydraco*.

DISCOVERY PROFILE

Name	*Eoraptor lunensis*
Discovered	Ischigualasto Provincial Park, Argentina, by Ricardo Martínez, 1991
Described	Paul Sereno and colleagues, 1993
Importance	One of the oldest and most primitive dinosaurs
Classification	Saurischia

Meat-eating bull
and its cousins

Ever since British paleontologist Arthur Smith Woodward described some incomplete theropod jaws from Patagonia that he named *Genyodectes serus* in 1901, it was assumed that South America's Cretaceous theropods were close relatives of Europe's *Megalosaurus* or North America's *Allosaurus*. Various other fragments found during the following decades did little to change this picture, but it did change radically in the 1980s.

Above When seen from the front like this, it is obvious that the horns of *Carnotaurus* were stout and thick. We can only guess what their function was. Roles in combat, display, and even predation have been suggested.

DISCOVERY PROFILE

Name	*Carnotaurus sastrei*
Discovered	Bajada Moreno, Chubut, Argentina, by members of the 8th Paleontological Expedition to Patagonia, 1984
Described	José Bonaparte, 1985
Importance	First well-represented abelisaurid, and one of the most bizarre theropods
Classification	Saurischia, Theropoda, Ceratosauria, Abelisauridae

During a paleontological expedition to the province of Chubut in Argentina, the skeleton of an unfamiliar large theropod was discovered embedded within a very hard rocky nodule at Estancia Pocha Sastre, near Bajada Moreno. It was fully articulated and excellently preserved, although the feet, the lower bones of the legs, and part of the tail were missing due to erosion. In 1985, Argentinean paleontologist José Bonaparte published a brief article in which he named this theropod *Carnotaurus sastrei,* which means "Sastre's meat-eating bull."

Carnotaurus was exciting for several reasons. Its strangely short-faced, tall, narrow, horned skull is unlike that of any other theropod; it raises many questions about the animal's behavior and lifestyle. Its skull bones have a corrugated surface texture. The opening

behind the eye socket (the laterotemporal fenestra) is proportionally huge; and the eye socket itself is tall, narrow, and constricted about halfway up by a large projection from the postorbital bone (a similar constricted eye socket is seen in tyrannosaurs).

Unusual vertebrae and arms

A full description of *Carnotaurus* was published in 1990, and it showed that the animal has extreme features elsewhere in its skeleton too. The shape of the vertebrae show that *Carnotaurus* had an unusual flattened back, in contrast to many theropods, in which a low bony ridge runs along the midline of the body. Its arms and hands are particularly bizarre. The upper arm bone, or humerus, is long and straight, but the lower arm bones— the radius and ulna—are very short and strangely oriented: the radius is located on the outside of the arm instead of at the front. Due to the way in which the radius is positioned, the palm of the hand must have faced forward. Bonaparte and colleagues illustrated the hand with four fingers, showing the fourth as a backward-pointing spike, and the first three as short, blunt, clawless stumps.

These details are strange indeed, and they were portrayed inaccurately in most reconstructions. Later work showed that the forelimb nerves were very small, which implied

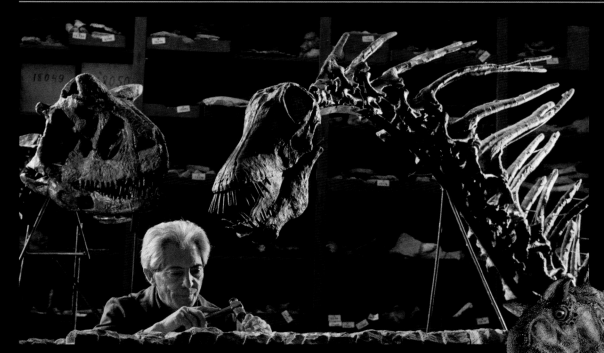

that the arms were weak—in fact, the elbow was immobile. However, a large, rounded head on the humerus enabled the entire arm to move with greater freedom than was usual for theropods, so the arm must have had some function.

Preserved skin

Carnotaurus was also very special in preserving skin impressions. Most of the animal's surface was covered by small, non-overlapping scales of the sort well known for dinosaurs, but scattered among these were larger, conical scutes that were probably arranged in rows. Unfortunately, skin impressions are so rare for large theropods that we do not know if such details were typical.

Another new large South American theropod was announced in 1985: *Abelisaurus comahuensis*. Known only from its skull, *Abelisaurus* was thought to be distinct enough to deserve its own new theropod family, the Abelisauridae. *Carnotaurus* shared several features with *Abelisaurus* that were not seen in other theropods and thus was placed within the same group. Here was an entirely new collection of large theropods, apparently unique to South America. These dinosaurs seemed to have pursued their own evolutionary trajectory for most of the Cretaceous. Other abelisaurids were soon described from the Cretaceous rocks of South America, and eventually it was revealed that these archaic theropods had been an important presence across the Gondwanan continents.

PICTURING THE HORNS

José Bonaparte's original 1985 article provided scant information about *Carnotaurus* and did not describe the exact form of its horns. As a result, artistic reconstructions that appeared in the next few years were substantially inaccurate. One painting published in 1989 showed the horns projecting straight up from above its eyes, and a well-known reconstruction produced in 1988 illustrated *Carnotaurus* with flattened, wing-like horns. In fact, the horns are thick from top to bottom and shaped more like blunt knobs than horns or wings. In the absence of more specimens, it is difficult to evaluate the function of these horns. As with the other horns and horn-like structures seen in dinosaurs, they may have been used in display, or in fighting, but such ideas remain entirely speculative.

133

The South American sauropod explosion

By the late 19th century, paleontologists were aware that sauropods inhabited South America during the Cretaceous, but few remains were known. A few vertebrae were reported during the 1890s, and a complete forelimb (named *Argyrosaurus superbus*) was described in 1893. Additional finds were described in the 1920s, when Friedrich von Huene named *Antarctosaurus*. Much later, in 1979, Jurassic sauropods were described from Patagonia. These seemed to be cetiosaur-type animals similar to those known from Europe. However, detailed information about Cretaceous sauropods was still thin on the ground.

Above *Amargasaurus* was the first dicraeosaurid sauropod to be reported from outside Africa. Its tall, bony spikes make it one of the most unusual of all sauropods.

DISCOVERY PROFILE

Name	*Amargasaurus cazaui*
Discovered	La Amarga, Neuquén Province, Argentina, by members of 8th Paleontological Expedition to Patagonia, 1984
Described	By Leonardo Salgado and José Bonaparte, 1991
Importance	A unique spike-necked sauropod, one of the best-known South American diplodocoids
Classification	Saurischia, Sauropodomorpha, Sauropoda, Diplodocoidea, Dicraeosauridae

During the 1980s, 1990s, and beyond, South America proved to be the most important continent for sauropod discoveries. A group of new researchers led by Argentinean paleontologist José Bonaparte revealed to the world a succession of amazing new sauropods.

Among the first was little *Amargasaurus cazaui* from the La Amarga Formation of Argentina's Neuquén Province, described by Leonardo Salgado and Bonaparte in 1991. For a sauropod, "little" indicates a length of about 30 ft (9 m). *Amargasaurus* was clearly a close relative of *Dicraeosaurus*, a short-necked diplodocoid from Tendaguru. Like other diplodocoids, *Amargasaurus* has proportionally short arms and a long skull in which structures normally positioned at the very back had rotated forward to lie underneath the eye socket. As in *Dicraeosaurus*, the spines on the tops of its vertebrae are very tall. What makes *Amargasaurus* strikingly different, however, are the tremendously long forked spines that grew up from the tops of its neck vertebrae.

Spine function

What role did these spines play? They were originally thought to have been incorporated into twinned skin sails that ran along the length of the neck. Perhaps these sails were used in sexual and social display, or in thermoregulation. However, in 1994 it was argued that the pointed tips of the spines, combined with the awkwardness of having paired skin sails on a long neck, instead suggested that the spines were sheathed in horn, and that they projected from the upper surface of the neck as independent spines. Ultimately, it is not possible to be sure either way, and we remain uncertain how *Amargasaurus* looked when alive.

The titanosaurs

Other new Cretaceous South American sauropods, although not as bizarre, were no less interesting. In 1980, Bonaparte and Jaime Powell named *Saltasaurus loricatus*, a mid-sized sauropod in which the back and sides are covered with small armor plates. Unlike *Amargasaurus*, *Saltasaurus* was a titanosaur. Titanosaurs were mainly Cretaceous sauropods that have distinctive tail vertebrae with convex posterior surfaces. During the 1980s and early 1990s, it was thought that these dinosaurs were close relatives of the diplodocoids, a view that owed itself (in part) to the discovery of slim, pencil-like teeth in some titanosaurs. Later work showed that titanosaurs are in fact more similar to *Camarasaurus* and the brachiosaurs. *Saltasaurus* was the first of many South American titanosaurs.

NOT SO STRANGE

The discovery of *Amargasaurus* and of the new titanosaurs, together with the recent recognition of the abelisaurid theropods, reinforced the idea that South America's Cretaceous dinosaurs were somehow different from those elsewhere, and that these were "strange" species that had evolved in isolation from those of northern continents. But later it was argued that the South American Cretaceous dinosaurs mostly belonged to groups that were widespread during the Jurassic. If anything, it was the Cretaceous dinosaurs of western North America and eastern Asia—the giant tyrannosaurs, horned dinosaurs, and elaborately crested lambeosaurine hadrosaurs—that were the radically strange ones.

In 1991, Jorge Calvo and Bonaparte named *Andesaurus delgadoi*. It came from Neuquén and was excavated in 1987, but unlike *Saltasaurus* and other advanced titanosaurs, it has peg-and-socket type processes in its vertebrae (called the hyposphene-hypantrum system), and thus seemed to be a primitive member of the group. In 1993, a second primitive form was named *Argentinosaurus huinculensis* by Bonaparte and Rodolfo Coria. The remarkable thing about *Argentinosaurus* was its size: with vertebrae more than 5 ft (1.5 m) tall and a thigh bone almost 8 ft (2.4 m) long, its overall length was estimated at almost 100 ft (30 m) and its weight at 90–110 tons (80–100 tonnes).

Above *Argentinosaurus* lived alongside a gigantic allosauroid theropod, *Mapusaurus*, which is shown here ambushing a juvenile, while the rest of the herd moves on. The latter, a close relative of *Giganotosaurus*, was discovered in 1997 but not named until 2006.

135

Giant dromaeosaurs

When the Cretaceous dromaeosaur *Deinonychus* was described from Montana in the late 1960s, it was widely hailed as one of the most dangerous animals ever to have lived. It boasts serrated teeth and dextrous hands with wickedly sharp, strongly curved claws, and its powerful hind limbs have large, extendable sickle claws. So it is not hard to see why the announcement of a new theropod similar to *Deinonychus* but much larger caused great excitement in the paleontological community.

Above Before the discovery of *Utahraptor*, the largest known dromaeosaurs, such as *Deinonychus*, were 10 ft (3 m) long. This reconstructed *Utahraptor* skeleton is more than twice as large as that of *Deinonychus*.

DISCOVERY PROFILE

Name	*Utahraptor ostrommaysorum*
Discovered	Gaston Quarry Utah, U.S.A., by Carl Limoni, 1991
Described	By James Kirkland and colleagues, 1993
Importance	First giant dromaeosaur to be discovered
Classification	Saurischia, Theropoda, Coelurosauria, Maniraptora, Dromaeosauridae

The remains of the giant dromaeosaur were excavated by James Kirkland and his colleagues in 1991. The bones were discovered at two sites, both in eastern Utah, in rock of the Lower Cretaceous Cedar Mountain Formation. The sickle claw from the foot, almost 10 in (25 cm) in length along its curved upper edge, showed that this new animal was more than twice the size of *Deinonychus*. They called it *Utahraptor* (*see* box).

Articulated remains of *Utahraptor* have not been found, but enough material has been discovered to reveal what this dinosaur looked like. Bones from the tip of its snout show that its head is approximately rectangular and shaped much like that of *Deinonychus*. Hind limb bones show that its legs were stocky and muscular. A few vertebrae from the tail show that its tail anatomy is similar to that of *Deinonychus*, and it has the extremely long, bony rods typical of the group.

Sickle claws

Dromaeosaurs have particularly large, strongly curved hand claws. The hand claws thought to belong to *Utahraptor* appear to be proportionally even bigger and more strongly curved than those of other dromaeosaurs. These claws are also particularly thin, with a very narrow, blade-like edge. These features suggested to Kirkland and colleagues that *Utahraptor* might have used its hand claws as cutting weapons, and that it was perhaps in the habit of tearing at the sides of its prey, or of latching on to them while kicking at them with the sickle claw on the foot. These blade-like hand claws also implied, they felt, that *Utahraptor* was a specialized offshoot and not ancestral to later dromaeosaurs. However, it was later shown that the unusual "hand claws" of *Utahraptor* were not hand claws at all, but sickle claws from the feet.

Active hunter

Kirkland and his colleagues estimated the length of *Utahraptor* as 23 ft (7 m) and its weight as about 1,550 lb (700 kg). This made it a giant compared to other dromaeosaurs. They realized that *Utahraptor* was almost certainly a hunter of large dinosaurs and noted that, if it practiced any sort of social hunting strategy (as has been proposed for *Deinonychus*), then even the biggest sauropods might have been part of its diet.

However, evidence for the diet and behavior of *Utahraptor* is lacking. We can only speculate on what it may have eaten. The geological unit that yielded *Utahraptor*'s remains—the Cedar Mountain Formation—has produced many herbivorous dinosaur fossils, including those of ankylosaurs, iguanodontians, and spatulate-toothed sauropods, and all of these potentially may have been hunted by *Utahraptor*.

A global sensation

Utahraptor's debut in 1992 and 1993 caused a sensation in the world's media. The fact that dromaeosaurs made an appearance in the 1993 movie *Jurassic Park* perhaps made the discovery more newsworthy than was usual. The large dromaeosaurs that starred in *Jurassic Park* were likened by some authors to *Utahraptor*, although they were referred to as *Velociraptor* in the script. As it happens, these animals were in fact modeled on *Deinonychus*: Michael Crichton, the author of the novel on which the film is based, had chosen to follow Greg Paul's 1988 proposal that *Deinonychus* was similar enough to *Velociraptor* to belong in the same genus. Other theropod specialists did not agree with this classification, and Paul also eventually rejected the idea.

Below If giant dromaeosaurs hunted cooperatively as shown here, they may have been able to tackle giant prey such as this sauropod. At the moment, direct evidence on the behavior and diet of *Utahraptor* is unknown.

GRAMMATICAL ERROR

When they named *Utahraptor*, Kirkland and his colleagues decided to honour both John Ostrom, the describer of *Deinonychus* and a major player in the dinosaur "renaissance" of the 1960s and 70s, and Chris Mays of Dinamation International Corporation. Accordingly, the new species was published as *Utahraptor ostrommaysi*, in 1993. Unfortunately, this Latin form is singular, and because the name honours more than one person, it was corrected to *Utahraptor ostrommaysorum*.

Illustrating dinosaurs

During the past few decades, the science and art of bringing dinosaurs to life has become more precise and more rigorous than was traditionally the case. Artists previously relied on skeletons mounted in museums for guidance, and dinosaurs were assumed to have bloated, flabby bodies and small, shapeless limb muscles. A new, anatomically rigorous movement in dinosaur art was initiated during the 1980s by American artist Greg Paul. Paul argued that artists should produce their own accurately scaled and posed skeletal reconstructions before attempting to "bring a dinosaur to life." Many good paleo-artists have followed his lead. Reconstructions of dinosaurs continue to be produced in paint or pencil, but improvements in digital technology have resulted in a plethora of computer generated dinosaurs. Unfortunately, a vast amount of extremely inaccurate artwork remains, much of it created by artists who are completely unaware of the literature on dinosaur reconstruction.

Insets, top to bottom To reconstruct a dinosaur accurately, an artist has to begin with an accurately restored skeleton. Old illustrations and museum mounts often have many inaccuracies, so new, high-fidelity reconstructions (top) are required. Based on what we know of living reptiles, the musculature of dinosaurs can be accurately reconstructed, though there is often room for error and uncertainties always remain. Traditionally, the skin textures of dinosaurs have been guessed at. However, fossilized skin patches show us what some dinosaurs were like on the outside. Color will always remain complete guesswork, and it is certainly the least important part of a reconstruction.

Main image Before the dinosaur renaissance, dinosaurs were typically illustrated as rather shapeless animals that lacked obvious muscles. They were often shown as dull, tail-dragging creatures, and anatomical errors abounded.

New discoveries and technologies have allowed a new generation of illustrators to paint vivid, exciting portrayals of the Mesozoic world. In this mixed-media piece by artist Luis Rey, the gigantic oviraptorosaur *Gigantoraptor* defends its nesting grounds from the tyrannosauroid *Alectrosaurus*.

Multi-toothed ostrich dinosaur

Ostrich dinosaurs, properly termed ornithomimosaurs, were known from the fossil record of North America since the 1890s, and Asian members of the group were discovered during the 1920s. These early finds, all from the Upper Cretaceous, are species with beaked, toothless skulls. Their eye sockets are enormous, and, like ostriches, they have long, slender necks, and long, powerful hind limbs that suggest they were fast runners.

Below Unlike other ostrich dinosaurs, *Pelecanimimus* had a very long, shallow snout and hundreds of tiny teeth. Quite what it did for a living nobody really knows.

Unlike ostriches, of course, ornithomimosaurs have a long, muscular tail, and their long arms end in slender, three-fingered hands. In contrast to other theropods (in which the thumb is usually shorter than the other fingers), the three fingers of the ornithomimosaur hand are nearly equal in length, and the claws are relatively straight, rather than curved.

New ornithomimosaurs

During the 1980s, a relatively primitive ornithomimosaur was discovered in Mongolia. It was named *Harpymimus* in 1984, and it differs from previously known forms in having about 20 teeth in its lower jaws. *Harpymimus* also differs from more advanced ornithomimosaurs in its wider, shorter feet, and its thumb, which is shorter than the other two fingers. The species suggested, as did other ostrich dinosaur fossils, that the

DISCOVERY PROFILE

Name *Pelecanimimus polyodon*

Discovered Las Hoyas, Cuenca, Spain,
by A. Díaz Romeral, A. Fregenal,
and N. Meléndez, 1993

Described By Bernandino Pérez-Moreno and
colleagues, 1994

Importance First European ornithomimosaur; one
of most primitive members of its group

Classification Saurischia, Theropoda, Coelurosauria,
Ornithomimosauria

More than 150
teeth in lower jaw

More than 70 teeth
in upper jaw

Skin pouch
that may have
stored food

Above Unfortunately
the only known
Pelecanimimus
specimen consists of
the animal's front half
only. However, skin
impressions reveal
rare details of shape
and anatomy.

ornithomimosaur group had originated in
eastern Asia and was unique to this region and
to North America.

In 1993, a remarkable new theropod was
discovered at Las Hoyas in Spain. Las Hoyas was
already well known as a place where exquisitely
preserved plant and animal fossils had been
found, and during the late 1980s and early
1990s, the discovery of fossil birds made the site
of particular interest to dinosaur researchers.
The new theropod was an entirely new kind
of ornithomimosaur—the first ever from
Europe—and in 1994 it was described by Spanish
paleontologist Bernandino Pérez-Moreno
and colleagues. They named it *Pelecanimimus
polyodon*. Its long skull and the possible presence
of a throat-pouch reminded them of a pelican,
so the generic name they chose means "pelican
mimic." The presence of numerous teeth inspired
the specific name, which means "many-toothed."

Amazing number of teeth

The skull of *Pelecanimimus* is truly remarkable.
It is far longer and shallower than that of other
ostrich dinosaurs, and a small blunt horn
is present just in front of and above the eye
socket. Lining its jaws are an incredible 220
or so unserrated teeth, more than in any other
theropod. These teeth are not all identical: those
at the front of the upper jaw are D-shaped in
cross section; and those near the back of the upper
jaw are more blade-like than the lower jaw teeth.

Unfortunately, the single known
Pelecanimimus specimen preserves only the front
half of the body. Nothing is present further back
than the shoulder girdle and breastbone. The
breastbone, or sternum, is large and similar to
that of birds and bird-like maniraptorans, such
as dromaeosaurs. Like other ornithomimosaurs,
all three of the fingers in *Pelecanimimus*
are about equal in length, so the hand

presumably functioned as a hook-like structure.
Pelecanimimus also preserves soft tissue
impressions. Besides skin under
the lower jaw, there is what
appears to be a soft crest at
the top of the skull.

Possibility of feathers

By the time *Pelecanimimus*
was described, some
authorities thought that small
theropods might have feathers or feather-like
structures, although the evidence for this was
wanting. *Pelecanimimus* was reported to have
parallel fibers on its arms, so did it preserve
evidence of a feather-like covering? A proper
investigation of this suggestion was not
carried out, but evidence for fiber-like "proto-
feathers" soon came from elsewhere among the
theropods (*see* p.124).

As is the case with so many exciting dinosaur
finds, the original, brief article that appeared
on *Pelecanimimus* is, to date, all that has been
published on the animal. A detailed description
has yet to appear.

TOOTH FUNCTION

There are as yet no data showing us what
Pelecanimimus might have done with its
hundreds of teeth. Was it using them to catch
small animals, nibble at leaves, or filter prey from
water? Pérez-Moreno suggested that the many
teeth formed a single cutting edge that acted like
a beak's edge. If this is true, however, it seems
strange that ostrich dinosaurs switched from
having lots of teeth to none at all. Given that
ornithomimosaurs are famous for being mostly
toothless, it's ironic that the theropod with the
highest number of teeth belongs to this group.

Evolution in action

Many ceratopsids—the great horned dinosaurs of North America's Late Cretaceous—were named during the late 19th century and the early 20th century, but this momentous phase of discovery eventually slowed. In the second half of the century, paleontologists thought that perhaps there were no new ceratopsids left to find—*Pachyrhinosaurus*, named in 1950 (*see* pp.86–87), seemed to be the last of them. But a new cycle in ceratopsid research and discovery began in the 1990s.

Below *Einiosaurus* had a forward-pointing nose horn and long, backward-pointing spikes on its short frill. As in *Styracosaurus*, the frill ornaments most likely played a role in display.

John Horner and colleagues published the results of a study on the rocks and dinosaurs of the Upper Cretaceous of the Rocky Mountain region in 1992. Here, several different layers of sediment were deposited when a branch of the interior seaway (a shallow sea that divided North America in two during the Cretaceous) moved in across Montana and Alberta and later receded. This event was known as the Bearpaw Transgression,

and the marine rocks laid down during the sea's incursion were known as the Bearpaw Shale. Immediately below the Bearpaw Shale was the Two Medicine Formation, the Upper Cretaceous unit best known among dinosaur specialists for having yielded the "good mother hadrosaur" *Maiasaura* (*see* pp.100–101).

Evolutionary intermediates

In 1987, Horner and his coworkers discovered the remains of new horned dinosaurs in the upper layers of the Two Medicine Formation, and in their 1992 report they described how these represented *three* new species. All appeared to be new types of styracosaur, and hence part of the centrosaurine group

DISCOVERY PROFILE

Name	*Achelousaurus horneri*
Discovered	Glacier County, Montana, U.S.A., by Jack Horner and colleagues, 1987
Described	By Scott Sampson, 1995
Importance	One of only a few ceratopsids named in late 20th century; an important intermediate between *Pachyrhinosaurus* and other centrosaurines
Classification	Ornithischia, Marginocephalia, Ceratopsia, Ceratopsidae

STYRACOSAURUS

CENTROSAURUS

CENTROSAURINAE 1

2

EINIOSAURUS

3

ACHELOUSAURUS

4

P. LAKUSTAI

5

P. CANADENSIS

Right Like the larger, geologically younger, *Pachyrhinosaurus*, *Achelousaurus* had a thickened nasal boss instead of a horn. This was not really a new structure in evolutionary terms, but was most likely a strongly modified horn.

Above Anatomical features shared by *Achelousaurus* and *Pachyrhinosaurus* show that these two share an ancestor that was similar to *Einiosaurus*. This einiosaur-pachyrhinosaur clade was very closely related to another clade that included *Centrosaurus* and *Styracosaurus*.

of ceratopsids, with short, deep snouts, short frills, and no brow horns. These new ceratopsids were particularly exciting because they appeared to form a series of evolutionary intermediates between *Styracosaurus* (with its long nasal horn and long spikes on the frill) and *Pachyrhinosaurus* (with its nasal boss and shorter frill spikes).

Horner and his team also reported other new dinosaurs from the top of the Two Medicine Formation: a new lambeosaurine hadrosaur that seemed to be ancestral to *Hypacrosaurus*; a dome-skulled pachycephalosaur that was apparently close to the ancestry of *Pachycephalosaurus*; and a new tyrannosaurid that appeared intermediate between *Daspletosaurus* and *Tyrannosaurus*.

Different horns

The new Two Medicine centrosaurines were fully described and named by Scott Sampson in 1995. One species has a forward-curved nasal horn and was named *Einiosaurus procurvicornis*, which means "buffalo lizard with the forward-curving horn." Sampson also reported another Two Medicine chasmosaurine with a nasal horn, but, in contrast to *Einiosaurus*, the horn in this dinosaur is tall, slim, and points upward. Too little was known about this second species to name it, but Sampson suggested that it might have been the adult form of *Brachyceratops*, a centrosaurine named in 1914 for several juvenile specimens.

Finally, a pachyrhinosaur-like species was named *Achelousaurus horneri*. The generic name referred to Achelous, a Greek deity who battled Hercules by turning into a bull. To defeat him, Hercules removed one of his horns. The species name was given in recognition of Horner's extensive and innovative work on Two Medicine dinosaurs. In place of a nasal horn, *Achelousaurus* has a corrugated nasal boss. This structure is reminiscent of the massive boss that Langston described for *Pachyrhinosaurus* in 1950, but the boss of *Achelousaurus* is shorter, narrower, and less massive.

It seems that *Pachyrhinosaurus* was the last member of a lineage in which nasal horns have been substituted for gnarled snout-lumps. Two long spikes project from the rear margin of the frill of *Achelousaurus*, and unlike those of *Einiosaurus*, they curve outward and away from each other. Clearly, North America had not yet given up all of its ceratopsid secrets—and more discoveries followed.

HABITAT REDUCTION THEORY

The discoveries from the top of the Two Medicine Formation raised intriguing questions. Was it coincidental that so many evolutionary intermediates came from a time and a place in which an encroaching seaway was restricting the available habitat? Horner and his colleagues thought not. They proposed that a rapid reduction in Two Medicine dinosaur habitat forced small populations of the local dinosaurs to evolve rapidly.

Above The massive
reconstructed skull of the
north African allosauroid
Carcharodontosaurus
dwarfs that of a human.
Bony shelves project
out over its eyes and
its teeth are flattened
and dagger-like.

Southern continent
super-predators

DISCOVERY PROFILE

Name	*Giganotosaurus carolinii*
Discovered	Río Limay Formation, Neuquén Province, Argentina, by Rubén D. Carolini, 1993
Described	Rodolfo Coria and Leonardo Salgado, 1995
Importance	South America's largest theropod
Classification	Saurischia, Theropoda, Allosauroidea, Carcharodontosauridae

One of the few "facts" that everyone seems to know about dinosaurs is that *Tyrannosaurus rex* was the largest of the predatory theropods. Much to the delight and interest of the world's media, this was challenged during the 1990s due to the discovery and redescription, respectively, of two new giant theropods. Both were closely related, and both came from the Gondwanan continents.

The first of the two dinosaurs was a new species from the Upper Cretaceous Río Limay Formation of Neuquén Province in Argentina. It was reported to be the largest theropod ever found in the Southern Hemisphere, and was named *Giganotosaurus carolinii*. It is represented by a partial skull, vertebrae, the shoulder girdle, the pelvis, and a hind limb. It is clearly a robustly constructed, deep-skulled theropod similar in overall proportions to *Allosaurus*. Shelf-like bony flanges grow out from the skull above the eye socket and a low, triangular horn projects upward from in front of it. The tip of the lower jaw is deep, with a small, chin-like process. The back of the skull is tall but narrow. The animal's arms were not discovered, but were inferred to be small because the shoulder blade is small.

Size matters

Coria and Salgado estimated that *Giganotosaurus* was perhaps 41 ft (12.5 m) long and weighed 6.5–9 tons (6–8 tonnes), in which case it was similar in size to *Tyrannosaurus*. However, given that the lengths and weights of giant theropods are difficult to estimate, it is impossible to be confident that either *Giganotosaurus* or *Tyrannosaurus* was really bigger than the other.

A second *Giganotosaurus* specimen, which was no more than an incomplete dentary bone from the lower jaw, was described in 1998. It was found in 1987, with the result that *Giganotosaurus* was in fact discovered some years before the recovery of Carolini's specimen (which was found in 1993). This dentary is about 8 percent larger than the dentary of Carolini's specimen. Given that the partial skull of Carolini's specimen was about 6 ft (1.8 m) long, the 1987 specimen suggests a total skull length of 6 ft 5 in (1.95 m).

With the description of *Giganotosaurus*, a new group of theropods was added to the list of South American Cretaceous dinosaurs. Whereas the abelisaurids were archaic and not closely related to the megalosaurs and allosaurs of the northern continents, *Giganotosaurus* was a tetanuran, and later work showed that it was a member of the same group as *Allosaurus*, the Allosauroidea.

Carcharodontosaurus

A second giant, Gondwanan theropod was reported in 1996. Unlike *Giganotosaurus*, this was not a new dinosaur, but rather a new specimen of an old one: it was new material of the North African *Carcharodontosaurus saharicus*. This species was named in 1927 and was initially misclassified as a species of *Megalosaurus*. However, in 1931, Ernst Stromer gave this theropod its own genus, naming it for the superficial similarity its teeth had with those of *Carcharodon*, the great white shark. The new material, which included a partial skull, was discovered on a University of Chicago expedition led by Paul Sereno, in 1995. With a total length of just over 5 ft (1.5 m), this skull was huge. More importantly, it showed that *Carcharodontosaurus* was a close relative of *Giganotosaurus*.

Above When announced in 1995, *Giganotosaurus* was hailed as the largest theropod of all time. Paleontologists generally find "competitions" such as this rather irrelevant, but they do generate a lot of media interest.

147

Mole-armed
Monoykus

Perhaps the most bizarre of the many new dinosaurs named in the 1990s was the small, long-legged, bird-like Mongolian theropod *Mononykus olecranus*. Originally called *Mononychus* (a name that was given to a beetle in 1824), *Mononykus* proved highly controversial.

Below With long legs, a long neck, small head, and short, powerful arms, *Mononykus* and its relatives—the alvarezsaurids—were strange little dinosaurs.

Although *Mononykus* superficially resembles a tiny ostrich dinosaur, it is unusual in that its arms are short, equipped with massive muscle attachment sites, and end in a short, robust hand that sports a large, stout thumb claw. Tiny facets suggest that reduced second and third digits were still present. This highly unusual forelimb looks suited for digging, yet the long neck and legs seem incongruous with a burrowing lifestyle.

Mononykus also has a large, boat-shaped breastbone, or sternum, that sports a prominent keel along its lower surface. The pubis and ischium in the pelvis are both slender and, uniquely, in close contact for their entire length; and the hind limbs are long and slender. The skull is poorly known, but appears bird-like and lightly built, with at least some small, conical-crowned teeth in the jaws. All in all, *Mononykus* displays a strange mixture of features.

What type of theropod?

The greatest controversy that surrounded *Mononykus* concerned its evolutionary relationships. What sort of theropod was it? Its describers proposed that it was a highly specialized, flightless bird, and a bird that was more closely related to modern birds that *Archaeopteryx* was. They coined the new name Metornithes for the clade that included

Mononykus and all other birds. Because *Mononykus* was regarded as a bird, it was reconstructed as feathered.

However, with its long tail skeleton and highly specialized arms, *Mononykus* would have to be a strongly modified bird indeed, and this proposal of avian status was soon challenged. Several paleontologists argued that *Mononykus* was simply too different from birds to be classified among them; they argued that its supposedly bird-like features (which include the large, keeled breastbone and fused wrist bones) are evolutionary consequences of its digging lifestyle, and only superficially similar to the features of true birds. Later work showed that *Mononykus* was not a member of the clade that includes *Archaeopteryx* and modern birds, but that its affinities lay elsewhere within the theropod group Maniraptora. Some experts even argued that *Mononykus* was not a maniraptoran at all, but a close relative of the ornithomimosaurs.

The alvarezsaurids

Mononykus was originally regarded as the only known member of its lineage. But in 1996, a second species was named: *Patagonykus puertai*, from the Upper Cretaceous of Neuquén Province, Argentina. Named by Argentinean paleontologist Fernando Novas, *Patagonykus* resembles *Mononykus* (it has similar arms, for example) but was about twice the size. Novas was also able to show that another Argentinean theropod, *Alvarezsaurus calvoi* (named by José Bonaparte in 1991), was a third member of the group. Bonaparte had published the name Alvarezsauridae for the group that included *Alvarezsaurus*, and as a result these strange little dinosaurs are known as alvarezsaurids today.

More tiny dinosaurs

Additional alvarezsaurid species were described later. A tiny Mongolian species, *Parvicursor remotus*, was named in 1996, and a second one, *Shuvuuia deserti*, in 1998. *Shuvuuia* includes a complete skull. In some ways, this skull is bird-like, with a flexible zone at the base of the rostrum, a reduced bony bar behind the eye socket, and a particularly large opening for the spinal cord. In other ways, it is very different, with nostrils positioned at the tip of the snout and numerous small, unserrated teeth lining the sides of the jaws. *Shuvuuia* is also interesting because its hand preserves small second and third fingers, both of which have long, straight claws. A North American alvarezsaurid was reported in 1998 (though its remains were too incomplete to warrant a name), and later work determined the presence of these dinosaurs in Europe and resulted in the naming of yet more new species from Asia and South America.

Above The tiny, bird-like skull of *Shuvuuia* has a very simple, elongated lower jaw and numerous tiny teeth. These are probably adaptations for a diet of small insects. Its skull was less than 3 in (8 cm) long, and the whole animal was only 2 ft (60 cm) in length.

Left It seems that all alvarezsaurids were feathered. However, complex, vaned feathers like those seen on the wings and tails of modern birds (and reconstructed here in *Shuvuuia*) were probably not present in the group.

149

Antarctica's "frozen crested" dinosaur

Throughout the Mesozoic Era, Antarctica was a tropical land that must have teemed with plant and animal life. Yet the rich fossil record that resulted is almost completely unobtainable today. Occasionally, however, a few places on the frozen continent have given us vital glimpses of the area's extinct animal life.

William Hammer and William Hickerson reported in 1992 the discovery of new Antarctic dinosaurs. They were collected during an expedition to Mount Kirkpatrick in the Beardmore Glacier Region in the Central Transantarctic Mountains. Here, an exposed rock unit called the Falla Formation revealed more than 60 bones, including the remains of three dinosaurs.

A prosauropod was represented by foot bones and a small theropod by teeth. Most important of all was the discovery of about 50 percent of the skeleton of a new large theropod, the partial skull of which was about 22 in (56 cm) long. These were not the first dinosaur fossils to be discovered in Antarctica: bones from an ankylosaur were described in 1987; and scraps from an ornithopod were published in 1991. The new theropod, however, was substantially well preserved in comparison to these early finds and was clearly something very special.

DISCOVERY PROFILE

Name	*Cryolophosaurus ellioti*
Discovered	Mount Kirkpatrick, Beardmore Glacier Region, Antarctica, by D. Elliot and others, 1990–91
Described	William Hammer and William Hickerson, 1994
Importance	The first well-preserved Antarctic dinosaur
Classification	Saurischia, Theropoda, Coelophysoidea, Dilophosauridae

Above The weird bony crest on the top of the cryolophosaur skull almost defies description. Bony crests were present elsewhere among theropods, but none resembled this.

Right The only known specimen of *Cryolophosaurus* (foreground) was discovered alongside several other dinosaur remains. The sauropodomorph *Glacialisaurus*, shown here in the background, was named in 2007.

Twin crests

Bizarrely, the Antarctic theropod was said to have a "large, furrowed crest that rises perpendicular to the long axis of the skull." Frustratingly, no pictures were provided, but in 1994 Hammer and Hickerson published a description of this remarkable theropod. They named it *Cryolophosaurus ellioti*, meaning "Elliot's frozen crested lizard." The deep and narrow skull has twinned crests on its upper surface. These crests superficially resemble the halves of a seashell, in which the concave surfaces face forward. The front of the skull has been removed by glacial erosion, but the skeleton also includes vertebrae and bones from the pelvis and hind limb.

Hammer and Hickerson thought that *Cryolophosaurus* was an allosaur-like theropod, and they made some comparisons between it and *Piatnitzkysaurus*, a megalosaur-like theropod from Patagonia. Their article includes a reconstruction showing the animal as it might have looked in life; it made *Cryolophosaurus* look much like *Allosaurus*, bar the unusual crest. In contrast to *Allosaurus* and related theropods, however, *Cryolophosaurus* was from the Early Jurassic and thus surprisingly old. Most other theropods from this time, such as *Dilophosaurus*, were more archaic.

Early Jurassic origins

The identification of *Cryolophosaurus* as a member of the Allosauroidea was supported by theropod workers during the 1990s. Its Early

Jurassic age was important, for it indicated that allosaur-type theropods originated much earlier than was thought. In turn, this implied that other groups of tetanuran theropods—including spinosaurs and coelurosaurs—appeared in the Early Jurassic, too. These conclusions may have been premature.

A full description of *Cryolophosaurus* was published in 2007. This work, produced by Nathan Smith and colleagues, showed that *Cryolophosaurus* was not an allosaur-like dinosaur as originally thought. In fact, it shares a list of skull characters with a group of Early Jurassic theropods that includes *Dilophosaurus* and *Dracovenator* from South Africa. Rather than being robustly built, like an allosaur, *Cryolophosaurus* was far slimmer, more lightly built, and had a shallower snout.

DEPICTING THE CRESTS

In the years following Hammer and Hickerson's paper on *Cryolophosaurus*, artists of prehistoric life struggled to depict the remarkable crest of this dinosaur accurately. Furrows along its length, breakage, and partial distortion (the crest was pushed downward on one side) all hampered interpretation, and some illustrations made the dinosaur look as if the crest was composed of several parallel, finger-like struts. Since large predatory theropods obviously had to tackle and subdue prey with their powerful jaws, it seems bizarre and counter-intuitive that they might have evolved elaborate bony head crests, especially fragile ones. Yet this is clearly what happened. *Dilophosaurus* from Arizona had already shown that some theropods have unusual head crests, and later on additional crested theropods were unveiled. Whatever role the crests played in the lives of these dinosaurs, that function was important in evolutionary terms.

Above Until recently, *Cryolophosaurus* was thought to be a close relative of such theropods as *Yangchuanosaurus* and *Allosaurus*. This reconstruction, which shows a deep, allosaur-like skull, is based on this idea.

151

Europe's
new hunter

Below One *Neovenator* specimen is beautifully preserved, and many regions of its skeleton are complete. Here, paleontologist Steve Hutt reconstructs the foot and skull.

The discovery of *Baryonyx* in 1983 (*see* pp.116–117) showed that the Lower Cretaceous Wealden rocks of southern England could still reveal new kinds of dinosaurs, despite being among the best known and most thoroughly studied in the world. In fact, another entirely new large theropod was found in these rocks in 1978, but it was not described until 1996. Unlike *Baryonyx*, it was discovered offshore on the Isle of Wight.

DISCOVERY PROFILE

Name	*Neovenator salerii*
Discovered	Grange Chine, Isle of Wight, England, by the Henwood family and others, in 1978
Described	Stephen Hutt and colleagues, 1996
Importance	One of Europe's most complete large theropods, and the oldest and most primitive carcharodontosaurid
Classification	Saurischia, Theropoda, Allosauroidea, Carcharodontosauridae

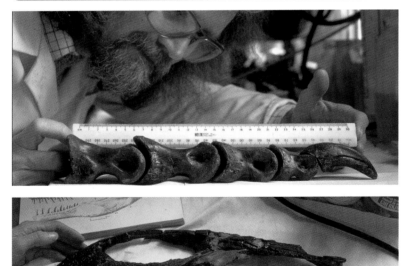

The new theropod was discovered by the Henwood family during a summer holiday, and it was the most complete Isle of Wight theropod yet found. Excavation carried out in following years gradually revealed more and more of its skeleton. Eventually, skull material, vertebrae, and bones of the shoulder girdles, pelvis, and hind limbs were recovered. A second specimen was discovered at a nearby site in 1987. This was originally thought to represent another new type of theropod, because the pair seemed to have differently shaped pubic bones: the pubis of the 1978 theropod has a massive, long "boot" at its end, whereas that of the 1987 one has a far shorter boot. To confuse matters further, the bones of the 1978 theropod were jumbled up with those of an iguanodontian.

By the 1990s, Stephen Hutt of the Museum Of Isle Of Wight Geology was able to show that both theropods were the same species. The short-booted "pubis" of the 1987 specimen turned out to be the ischium, another pelvic bone. Together with colleagues, Hutt named the species *Neovenator salerii* in 1996. The generic name means "new hunter" and was presumably inspired by *Afrovenator*, a then brand-new African theropod named by Paul Sereno and colleagues. The species name honored Mr. Salero, owner of the land on which the first specimen was discovered.

Related to *Allosaurus*

Neovenator was a large predator, about 25 ft (7.5 m) long, and probably similar to its older relative *Allosaurus*. It was certainly a member of the same major group—the Allosauroidea—but beyond that there was uncertainty about its relationships. Hutt and colleagues noted resemblances to both *Allosaurus* and the mostly Asian sinraptorids (the group that includes *Yangchuanosaurus* from China).

The front half of *Neovenator*'s skull was discovered and its bones were immaculately preserved, like those from the rest of the skeleton. Like *Allosaurus*, *Neovenator* is unusual in having five teeth in each premaxilla (the paired premaxillae are the two bones that form the tip of the snout). Other theropods generally have four premaxillary teeth. Its nostril opening is proportionally very large, and in later work Hutt showed that twin ridges run along the upper surface of its snout, terminating above the eyes and forming a V-shaped valley along the top of the skull. Many reconstructions of *Neovenator*, most of which make it look essentially the same as *Allosaurus*, depict it without these ridges, but they would have been obvious in life and probably enlarged by horny sheaths.

The vertebral column of *Neovenator* is more pneumatic than that of most other allosauroids, with openings for air sacs present on all of the vertebrae from the animal's back. This feature hints at a relationship with the gigantic carcharodontosaurids of North America and Gondwana, as did a few other details. Its forelimbs were unknown.

SICKLY SPECIMEN

The 1978 individual of *Neovenator* was not in perfect health before its death. Several of its bones were broken and had not fully healed, and a large unnatural lump was present on one of its shin bones. Distinctive pits on the articular surfaces of some of its toe bones hinted at some kind of bone disease, and one of its toe claws was damaged and blunted by some sort of traumatic injury.

The Wessex Formation

Neovenator was discovered in a layer of Wealden sediment called the Wessex Formation, which was already well known as one of the richest dinosaur-bearing sediments in Europe. The small ornithopod *Hypsilophodon* and its larger relative *Iguanodon* are known from here, as is the ankylosaur *Polacanthus*, the alleged pachycephalosaur *Yaverlandia* (it is probably, in fact, an unusual theropod), and a list of poorly known sauropods and small theropods. *Baryonyx* was found in the Wessex Formation in 1996, as was *Eotyrannus* in 1997 (*see* pp.164–165).

Below In life, *Neovenator* had tall, twinned ridges running along the top of its skull. No direct information is available on its diet, but it may have hunted small ornithischians such as these *Hypsilophodon*.

153

The Zuni Basin horned dinosaur

Since the discovery of *Psittacosaurus* and *Protoceratops* in Mongolia during the 1920s, paleontologists had a pretty good understanding of the early history of the horned dinosaurs—the ceratopsians. However, many mysteries remained, and one of the greatest concerned the origins of the best known ceratopsians, the ceratopsids.

DISCOVERY PROFILE

Name *Zuniceratops christopheri*

Discovered Zuni Basin, New Mexico, U.S.A., by Christopher Wolfe and family, 1996

Described Douglas Wolfe and James Kirkland, 1998

Importance A "missing link" from a previously unknown region of the ceratopsian family tree; one of very few North American dinosaurs from the "middle" Cretaceous

Classification Ornithischia, Marginocephalia, Ceratopsia

Facing page
Paleontologists assume that, like other horned dinosaurs, *Zuniceratops* used its bony frill and long brow horns in both display and combat.

Right *Zuniceratops* is an excellent transitional fossil. Many of its features appear intermediate between those of small, basal ceratopsians such as *Protoceratops* and the large, rhino-like ceratopsids such as *Triceratops*.

Most primitive ceratopsians are Asian, but a few are North American. One of these, *Montanoceratops* from Montana and Alberta, was thought by some experts in the 1990s to be more closely related to ceratopsids than to other ceratopsians. If true, this was evidence that ceratopsids had evolved in North America from a North American ancestor. However, even *Montanoceratops* is substantially different from the ceratopsids in many of its features, and it seemed that there was still a major gap in the fossil record between the ceratopsids and their smaller, more primitive relatives.

Then, in 1998, Douglas Wolfe and James Kirkland described a new North American ceratopsian. They named it *Zuniceratops* for the Zuni Basin of eastern Arizona and western New Mexico (the basin is itself named after the Zuni people). Wolfe and his young son Christopher had discovered the specimen in the rocks of the Moreno Hill Formation in New Mexico two years earlier. Before this discovery, dinosaurs had never been reported from these rocks, yet the Wolfes found evidence for several other species too, including hadrosaurs and theropods.

An intermediate species

The original *Zuniceratops* specimen was discovered weathering out of the side of a slope, as is often the case with dinosaur skeletons. The remains of several other specimens were soon discovered. *Zuniceratops* was important for two reasons: it lived at a time in which dinosaurs are poorly known (the Turonian, in the middle of the Cretaceous); and it seemed to be an ideal intermediate between animals such as *Montanoceratops* and the ceratopsids.

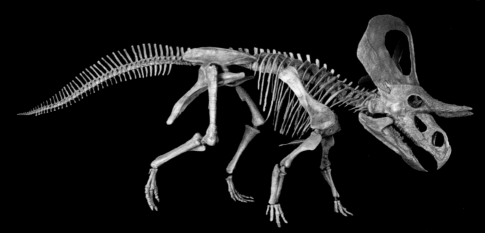

Zuniceratops is intermediate in size, with an estimated length of 10–11½ ft (3–3.5 m) and, like more primitive ceratopsians, is more lightly built than a ceratopsid. Although the teeth of ceratopsids are double-rooted, *Zuniceratops* shows the primitive condition of single-rooted teeth. Moreover, ceratopsids have both a nasal horn and brow horns, whereas *Zuniceratops* has only brow horns.

Brow horns

The long brow horns of *Zuniceratops* are in fact the most important fact about its anatomy. They gently curve upward along their length and are slightly compressed from side to side. Long brow horns such as these are not seen in more primitive ceratopsians, and were thought to have been restricted to the ceratopsids.

Given that other details of *Zuniceratops'* anatomy showed that it was not a part of Ceratopsidae, here was evidence that all ceratopsids evolved from ancestors with long brow horns. The implication is that centrosaurine horned dinosaurs (the group that includes *Styracosaurus*, *Pachyrhinosaurus*, and their relatives) evolved from an ancestor with long brow horns, but that later members of the group had lost them. *Zuniceratops* was also ceratopsid-like in having a long frill. In order to recognize the fact that *Zuniceratops* is closer to ceratopsids that it is to other ceratopsians, Wolfe and Kirkland created a new name, Ceratopsomorpha, for the "brow-horned ceratopsian" clade.

A MYSTERY SOLVED

One of the most peculiar bones attributed to *Zuniceratops* was a strange squamosal (the bone that normally forms the rear corner of the skull, and in ceratopsians forms the long side margin of the frill). The *Zuniceratops* squamosal looked bizarre, with a projecting, blunt-tipped rod and a serrated edge on its rear border. This odd bone was later discovered to be an ischium (a pelvic bone) from a theropod, and not a squamosal at all. One of the theropods discovered in the Moreno Hill Formation was a therizinosauroid—the first definite member of the group to be discovered in North America—and here was its ischium. The Moreno Hill therizinosauroid was named *Nothronychus* by Kirkland and Wolfe in 2001.

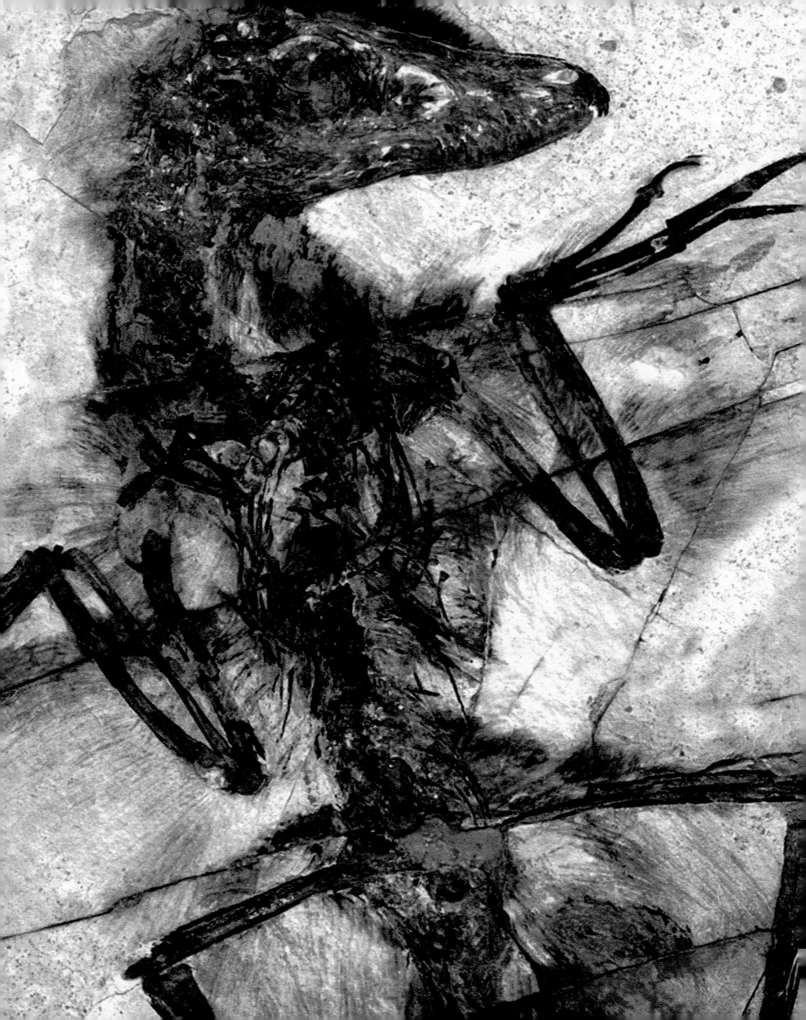

21st-century Dinosaurs

During the early 21st century, Chinese discoveries dominated the field of dinosaur research. Numerous feathered theropods were discovered, some of which, such as the dromaeosaur *Microraptor*, belonged to groups long suspected to be feathered. Others were more surprising, such as the small, feathered tyrannosaur *Dilong*. Thanks to technological advances, the study of dinosaur biology became increasingly sophisticated at the same time, and CT-scanning and digital modeling became a routine part of research on dinosaur biology.

Facing page Following the Chinese discoveries of the 1990s, massive international interest and increased searching resulted in the post-2000 naming of many new feathered theropods. The discovery of this spectacular feathered dromaeosaur was announced in 2001.

Alberta
Montana
Wyoming
Utah
New Mexico
Texas
Oklahoma
Coahuila

Maranhão
Pernambuco

Minas Gerais

Rio Grande do Sul
Mendoza Province

Neuquén Province
Rio Negro Province

Chubut Province
Santa Cruz Province

The 21st century

Above During the early 21st century, new abelisaurid theropods were named from Argentina, India, and Niger. *Rajasaurus*, shown here, is from India.

Spectacular dinosaur discoveries continue to be made today. China and Argentina have produced a flurry of unusual new species, but the new generation of paleontologists specializing in dinosaurs have continued to describe new species from Europe, North America, and elsewhere. Since 2000, we have seen such wonders as the long-snouted giant maniraptoran *Austroraptor*, the strange little theropod *Masiakasaurus*, the abelisaurs *Rugops*, *Rajasaurus*, and *Skorpiovenator*, the gigantic sauropod *Turiasaurus*, the dwarf sauropod *Europasaurus*, and the basal ceratopsians *Liaoceratops* and *Yinlong*.

Mount Kirkpatrick

2000 Giant sauropod *Sauroposeidon* named from Oklahoma
2000 *Microraptor* is in print
2000 New giant hadrosaur *Charonosaurus* is named

2002 Bizarre little *Epidendrosaurus* is published

2003 New crested hadrosaur *Olorotitan* is named
2003 *Tyrannosaurus* specimen "Sue" described in full

158 **2000** | **2001** | **2002** | **2003** | **2004**

2001 *Masiakasaurus*, a small Madagascan theropod with weird teeth, is named
2001 Basal tyrannosaur *Eotyrannus* is described

2001 Tribute volume to John Ostrom is published
2001 Oviraptorosaurs *Khaan* and *Citipati* are named
2001 The first therizinosauroid from North America, *Nothronychus*, is named

2004 Small, feathered tyrannosaur *Dilong* is published
2004 The "sleeping theropod" *Mei* is described

Kundur

Dorset
East Sussex
Isle of Wight

Teruel
Guimarota
Lleida
Valencia

Xinjiang

MONGOLIA

Trans-Altai Gobi
Nei Mongol
Dornogov Aimag

Liaoning Province
Heilongjiang Province

Shanxi Province
Gansu Province
Hebei Province
Henan Province

Ténéré Desert

Gujarat

Guangxi Zhuang

Chaiyaphum Province

Lindi
Malawi
MADAGASCAR

Free State Province

he early 21st century saw the emergence of a biomechanics movement, stimulated by new techniques and technologies, that sought to test many of the old assumptions about dinosaur function and behavior. The bite strength, head posture, neck mobility, arm strength, and running ability of many dinosaurs have been analyzed. This work should be seen as part of a broader approach to form and function in animals. The general thinking during much of the 1990s was that scientists had nothing new to learn from anatomy, and that genetics remained the only biological frontier, but this view has proved mistaken: it turns out that plenty of very basic questions about the anatomy of living animals have never been analyzed—and dinosaurs are at the heart of this anatomical revolution.

Theropod insights

The discovery of *Sinosauropteryx*, *Caudipteryx*, and *Protarchaeopteryx* in the 1990s proved that small theropods were feathered (or had, at least, quill-like "proto-feathers"). Beautifully preserved, complete or near-complete small theropods continued to be found during the following decade. Most of these fossils came from the Lower Cretaceous Yixian Formation in China's Liaoning Province, and most were described by Xu Xing and his colleagues, of the Institute of Vertebrate Paleontology and Paleoanthropology.

2005 Primitive therizinosauroid *Falcarius* is described
2005 Embryos of *Massospondylus* are described

2007 A sauropodomorph from Antarctica: *Glacialisaurus* is named
2007 Giant oviraptorosaur *Gigantoraptor* is named
2007 New Canadian ceratopsid *Albertaceratops* is named

2007 Weird new sauropod *Xenoposeidon* is described
2007 *Eotriceratops*, another Canadian ceratopsid, is named

2004

2005

2006

2007

2006 Basal tyrannosaur *Guanlong* is named
2006 North American pachycephalosaur *Dracorex* is named

2004 *Rugops* the abelisaur is named

2008 Giant maniraptoran *Austroraptor* described from Argentina

The bizarre *Microraptor*—a small dromaeosaur with long feathers on the hind limbs as well as the forelimbs—was named in 2000; the tiny, long-fingered *Epidendrosaurus* saw print in 2002; poorly known *Yixianosaurus* was described in 2003; and a small troodontid preserved in a sleeping posture, *Mei long*, was named in 2004. Many other such dinosaurs were named as well, and new, feathered dromaeosaur, oviraptorosaur, and therizinosauroid specimens provided much new information on the anatomy and feathery covering of these animals.

Given their position in the theropod cladogram, several groups, including tyrannosauroids and ornithomimosaurs, were believed to have started their history with feathers. This hypothesis was confirmed in 2004 with the description of the basal tyrannosauroid *Dilong*, yet another new small theropod from Liaoning. The early history of tyrannosauroids had long been poorly known, but several primitive forms were described during the first decade of the 21st century. *Eotyrannus* was named from England in 2001, *Guanlong* from China in 2006, and a new, English species of *Stokesosaurus*—*S. clevelandi*—in 2008.

Prosauropod progress

New theropod discoveries routinely make the headlines due to the popularity of these dinosaurs with the public, but a substantial amount of work on a somewhat less charismatic group—the basal sauropodomorphs or "prosauropods"—has appeared since 2000. A prosauropod "research renaissance" is in progress, and in 2007 an important, multi-authored volume on the group appeared.

The renaissance involved not only descriptive work, but exciting investigations of dinosaurian biology, growth, and lifestyle. South African *Massospondylus* embryos, described in 2005, showed that juveniles were fundamentally different from adults, and analysis of the forelimb

anatomy of *Massospondylus* and *Plateosaurus* revealed that these animals could not use their forelimbs in regular walking, as was thought. A 2005 study on bone growth in *Plateosaurus* showed that different individuals grew at different rates. Some matured quickly, and others grew slowly, at rates comparable to those of living crocodiles.

Sauropod studies

The biology and diversity of the giant sauropodomorphs, the sauropods, with their column-like limbs, also became the focus of renewed paleobiological work. Questions about the evolution of the unusual limbs and hands of these giant dinosaurs were illuminated by the discovery of primitive forms such as *Antetonitrus*. Studies on growth rates deduced from the internal structure of bone confirmed that sauropods grew astonishingly quickly. Particularly interesting new sauropods described during the early 21st century include the giant brachiosaur *Sauroposeidon*, the short-necked diplodocoid *Brachytrachelopan*, and the highly atypical *Xenoposeidon*, currently known only from a single vertebra fundamentally distinct from those of all other sauropods. New discoveries also showed that titanosaurs, which are traditionally regarded as restricted mainly to the southern continents, also have a good Asian fossil record.

161

Oklahoma's giant brachiosaur

By the late 20th century, it was apparent that sauropods were important and diverse Cretaceous dinosaurs in South America. In North America, however, it seemed that following their heyday in the Late Jurassic, sauropods had become very rare on the continent and the group as a whole had dwindled to obscurity or near extinction.

Below Comparison with *Giraffatitan* suggests that *Sauroposeidon* has a neck 39 ft (12 m) long. Here, it is shown as being about 15 percent bigger than *Giraffatitan* in total size. This is a reasonable estimate, but it may not be accurate.

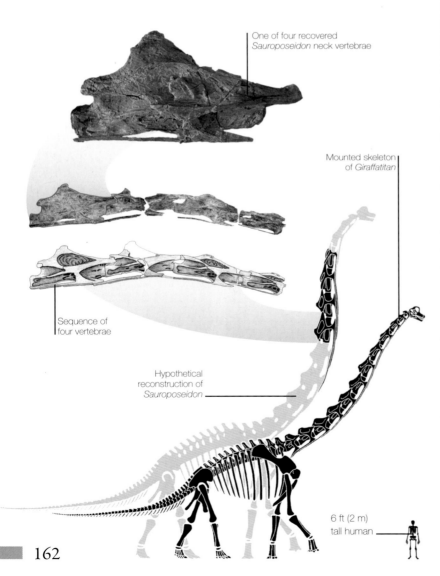

One of four recovered *Sauroposeidon* neck vertebrae

Mounted skeleton of *Giraffatitan*

Sequence of four vertebrae

Hypothetical reconstruction of *Sauroposeidon*

6 ft (2 m) tall human

Most North American sauropod remains were labeled either as *Astrodon johnstoni*, a species named in 1865 for teeth discovered in Maryland, or as *Pleurocoelus altus*, also from Maryland and named in 1888 for a few bones. However, it became clear that these labels obscured a greater diversity, and an impressive discovery made in Oklahoma in 1994 helped to show that the conventional view of North American Early Cretaceous sauropod diversity was woefully incorrect.

Immense vertebrae

The new find consisted of four huge neck vertebrae from a giant brachiosaur. The longest vertebra was 4 ft (1.2 m) long, which made it the longest complete vertebra ever found: the neck vertebrae of the diplodocoid *Supersaurus* (*see* pp.118–119) were longer, but the largest known specimen is missing its articulating processes, so its total length remains unknown.

The Oklahoman vertebrae are much like those of *Giraffatitan*, but longer and able to contain larger air sacs. In 2000, Mathew Wedel and colleagues based the new species *Sauroposeidon proteles* on the remains. Based on the size of its vertebrae, *Sauroposeidon* must have been about 92 ft (28 m) long with a neck about 39 ft (12 m) long. The complete neck probably included 13 vertebrae in total. Its name means "earthquake god perfected before the end," which reflects the idea that it was the culmination of brachiosaur evolution.

Apparent rarity

The discovery of *Sauroposeidon*—a new species not all that different from its Jurassic relatives—raised an interesting question: why were sauropods so much rarer in the Early Cretaceous of North America than in the Late Jurassic? In the Upper Jurassic rocks of the Morrison Formation, many sauropod species were known, and they were represented by hundreds of specimens. But in the Lower Cretaceous rocks of the continent, fragments were rare, and the number of species seemed to be much lower.

Some paleontologists suggested that an end-Jurassic extinction event had mainly removed sauropods from the North American fauna, leaving Cretaceous dinosaur communities dominated by ornithischians.

Diversity revealed

The apparent rarity of North American sauropods during the Early Cretaceous was illusory—we now know that diversity was high. The Morrison Formation covers a vast area, but North America's Lower Cretaceous rocks are less widespread and have not been examined as comprehensively. Moreover, the tradition of referring all North American Early Cretaceous sauropods to either *Astrodon* or *Pleurocoelus* brought about the self-fulfilling concept that diversity was low.

New specimens were found during the 1990s and beyond: *Sonorasaurus* was named from Arizona in 1998, and *Cedarosaurus* from Utah in 1999. After *Sauroposeidon* in 2000, *Venenosaurus* was named from Utah in 2001, and *Paluxysaurus* from Texas in 2007. Several other North American Early Cretaceous sauropods are currently in the process of being described. What is notable is that these sauropods are mostly brachiosaurs and titanosaurs, and not the diplodocoids or camarasaurs that are so abundant in the Morrison Formation. This suggests that sauropod faunas changed across the Jurassic–Cretaceous boundary, and it also seems that sauropods became less important ecologically as they are much less numerous.

Lower Cretaceous Europe also has a rich sauropod fauna, and again it seems that brachiosaurs and titanosaurs were important. In fact, some British fossils appear to represent brachiosaurs similar to *Sauroposeidon*.

Above Gigantic sauropod footprints such as this one from Texas—it is more than 3 ft (1 m) wide—are from the right time and place in geological terms to have been made by *Sauroposeidon*.

DISCOVERY PROFILE

Name	*Sauroposeidon proteles*
Discovered	Atoka County, Oklahoma, U.S.A., by Bobby Cross, 1994
Described	By Mathew Wedel and colleagues, 2000
Importance	The largest brachiosaur and one of the longest-necked sauropods
Classification	Saurischia, Sauropodomorpha, Sauropoda, Brachiosauridae

An unusual early tyrant

Despite a long history of detailed study, the Lower Cretaceous Wealden rocks of southern England surprised everyone during the late 20th century by revealing two new large theropods: *Baryonyx* in 1983, and *Neovenator* in 1978. In 1997, another surprise find was revealed. It was discovered by amateur collector Gavin Leng and, like *Neovenator*, it came from the Wessex Formation of the south coast of the Isle of Wight.

The new theropod is known from skull bones, parts of the forelimb, the hind limb, shoulder girdle, and vertebral column. It was medium-sized for a theropod: its bones suggest a total length of 10–13 ft (3–4 m). But the animal's most striking feature is its nasal bones: ordinarily in dinosaurs, these are two long, parallel, strip-like bones arranged along the upper surface of the snout, whereas in this species they are fused together and form a thick, corrugated "nasal unit" that has internal hollows and looks well suited for resisting considerable stresses.

Although several small theropods were named from the Wessex Formation during the 19th century, this animal was unlike any of them. It was clearly a new species, but what was it? Those fused, thickened nasal bones and other features indicated that this was a close relative of *Tyrannosaurus* and its kin, the tyrannosaurids, but that it was more primitive.

Early tyrant

A scientific paper on the new dinosaur was published by Stephen Hutt and colleagues in 2001. It was identified as a basal tyrannosauroid (Tyrannosauroidea is the more inclusive clade that contains Tyrannosauridae), and was named *Eotyrannus lengi*, which means "Leng's early tyrant." *Eotyrannus* lived alongside the small ornithopods *Hypsilophodon* and *Valdosaurus*, the ankylosaur *Polacanthus*, and a variety of large ornithopods and sauropods, but it presumably tried to avoid the attentions of the larger theropods *Neovenator* and *Baryonyx*.

As in tyrannosaurids, the premaxillary bones at the tip of the snout are short and show that *Eotyrannus* had a blunt, squared-off snout tip. The teeth in the premaxilla are shaped like an upside-down U when seen in cross-section, with the curved part of the U facing outward, and this is another feature also present in the Late Cretaceous tyrannosaurids.

Below Enough remains of *Eotyrannus* have been found to allow the creation of this reconstruction. Some of its bones, such as the dentary from the lower jaw and hand claw shown here, are incomplete but very well preserved.

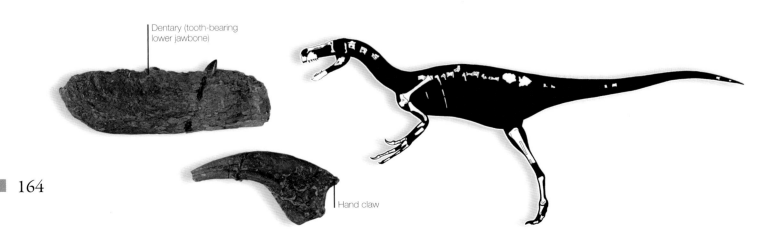

Dentary (tooth-bearing lower jawbone)

Hand claw

EVOLUTIONARY SIDE BRANCH

Recent work indicates that *Eotyrannus* was perhaps a peculiar side branch in tyrannosauroid evolution, and was probably not directly ancestral to the later tyrannosaurids. At the tip of each lower jaw bone, it has paired bony prongs that project upward along the midline, and the bones at the back of its lower jaw are unusual and unlike those of tyrannosaurids. It remains unknown what significance these features have for its way of life.

The arms and hands of *Eotyrannus* are relatively long, however, compared with those of tyrannosaurids. Its hands have three fingers, all of them clawed. These features suggest that *Eotyrannus* used its arms and hands to grab and restrain prey, but its skull and teeth show that it also had a very powerful, crushing bite. It seems that later tyrannosauroids—the tyrannosaurids and their close relatives—came to rely on the crushing bite almost entirely, eventually reducing the forelimbs in length and importance. Like its tyrannosaurid relatives, *Eotyrannus* has long, slender hind limb bones, and was presumably a fast runner.

Small beginnings

By 2001, it was widely agreed that tyrant dinosaurs are members of Coelurosauria (the theropod clade that includes birds and all the bird-like theropods). It follows that tyrannosauroids probably started their evolutionary history at small size. *Eotyrannus* was smaller than any tyrannosaurid, but it was still quite large. Furthermore, because the different parts of its vertebrae were not yet fused together (as they usually are in adult animals), the only known specimen was clearly not an adult at the time of death. An adult might therefore be somewhat larger.

The discovery of *Eotyrannus* hints at the possibility that tyrant dinosaurs have their roots in Europe, and not in eastern Asia or North America as paleontologists used to think, based on other fossil evidence. However, more primitive tyrannosauroids discovered in Asia suggest that, by the Early Cretaceous, basal tyrannosauroids were widespread in the northern continents. Given that even more primitive Jurassic tyrannosauroids occurred in Europe and North America, it may be that this combined region was where members of the group had their humble beginnings.

Below *Eotyrannus* was smaller and more lightly built than the tyrannosaurids of the Late Cretaceous. Even an adult was less than 23 ft (7 m) long, in contrast to 30 ft (9 m) or more for a typical tyrannosaurid. Its head was smaller proportionally: tyrannosaurids typically have heads about half the length of the body, but the head of *Eotyrannus* is only a third of body length; and *Eotyrannus* had long, three-fingered hands, rather than the stubby, two-fingered hands of tyrannosaurids.

DISCOVERY PROFILE

Name *Eotyrannus lengi*

Discovered Southwest coast of the Isle of Wight, England, by Gavin Leng, 1997

Described By Stephen Hutt and colleagues, 2001

Importance The first well-represented basal tyrannosauroid and the best tyrant dinosaur from Europe

Classification Saurischia, Theropoda, Coelurosauria, Tyrannosauroidea

165

Tiny long-fingered theropods

The discovery of therizinosauroids, abelisaurids, alvarezsaurids, and unusual sauropods, such as *Amargasaurus*, showed that dinosaurs evolved in weird and wonderful trajectories. In 2000, more exciting news came with the discovery of an extraordinary and entirely new kind of dinosaur: a tiny theropod with grasping feet and very unusual hands. In 2002, this bizarre little fossil was described and named, as was a second specimen of the same type of animal.

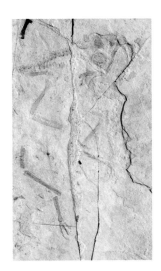

Above The tiny, partially disarticulated skeleton of *Epidendrosaurus*, shown here, is that of a juvenile. Adult specimens are currently unknown.

Right The only known *Epidexipteryx* specimen is well-preserved and near-complete. However, its hands are poorly preserved and some of its tail might be missing.

The first of the two fossils to be published was *Epidendrosaurus ningchengensis*, from the Daohugou Formation of Ningcheng County, northeast China. The age of the Daohugou Formation is controversial. It might be as old as Middle Jurassic or as young as Early Cretaceous; the most recent studies suggest an Early Cretaceous age. *Epidendrosaurus* was named by Fucheng Zhang and colleagues. It is tiny, measuring less than 8 in (20 cm) long, and has very long hands and a long tail (most of the tail bones have been lost and only an impression remains). The specimen appears to be a juvenile, possibly a very young hatchling.

Climbing wing

The second specimen comes from Liaoning Province, also in northeast China, and its skeleton is more complete than that of *Epidendrosaurus*. It was named *Scansoriopteryx heilmanni*. Its generic name means "climbing wing" and the specific name honors Gerhard Heilmann. Heilmann was a Danish artist and author, and his 1927 book *The Origin of Birds* was the standard work on the subject of bird evolution until Ostrom's work of the 1960s.

The *Scansoriopteryx* specimen preserves a far better skull than *Epidendrosaurus*. This is short, with enormous eye sockets and no evidence for teeth. The specimen also includes an articulated hand and a more complete pelvic girdle than

that of *Epidendrosaurus*. What makes the three-fingered hand so unusual is that the third finger is much longer than either the first or second. In theropods, the second finger is usually the longest.

In the foot, the first toe is positioned low down and close to the other three toes, and its position suggests that it may have been used to grasp twigs and branches when climbing. Unlike that of birds and dromaeosaurs, the pubic bone is not turned backward, but projects forward and downward in the manner typical for saurischians. Filament-like feathers appear to be preserved adjacent to the hand and elsewhere on the specimen, and what might be a small patch of scales is present near the end of the tail. This specimen also appears to be a hatchling.

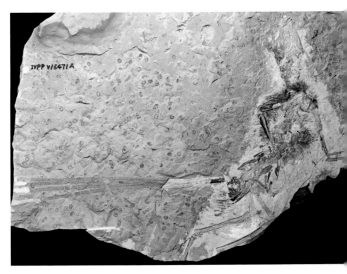

DISCOVERY PROFILE

Name	*Epidendrosaurus ningchengensis*
Discovered	Daohugou, Ningcheng County, northeastern China, *c.*2000
Described	By Fucheng Zhang and colleagues, 2002
Importance	The first truly tiny theropod to be discovered, perhaps one of the closest relatives of birds
Classification	Saurischia, Theropoda, Coelurosauria, Maniraptora, Scansoriopterygidae

NAMING ISSUE

Epidendrosaurus and *Scansoriopteryx* appear to be identical in most of their details, and it is now generally agreed that they are indeed the same animal. However, because both fossils were published at about the same time, it has proved difficult to establish which name is older and thus has technical priority. *Epidendrosaurus* has been used more frequently than *Scansoriopteryx*, but uncertainty still remains over which name is really the "correct" one.

Ribbon tail

A new *Epidendrosaurus*-like specimen was described in 2008. Named *Epidexipteryx hui*, this is also, like *Epidendrosaurus*, from the Daohugou Formation. It is a small animal too, but differs from *Epidendrosaurus* in having long, forward-pointing teeth at the front of its short, tall skull and in having a short tail. However, some of its tail vertebrae are missing and it may originally have had a long tail. What makes the specimen most remarkable is that four ribbon-like structures project from its tail vertebrae. Ribbon-like feathers are used during the display rituals of some birds, so it was suggested that these structures were also used for display in *Epidexipteryx*.

Placing the scansoriopterygids

Today, these unusual little theropods are referred to as the scansoriopterygids. They have several features that show they are part of Maniraptora. Some studies conclude that scansoriopterygids are more closely related to the clade that includes *Archaeopteryx* and modern birds than are other bird-like maniraptorans. However, scansoriopterygids resemble the members of certain other maniraptoran groups in some of their features, so it is possible that this position will change.

We still know very little about these animals. Given that all three specimens are juveniles or (in the case of *Epidexipteryx*) subadults, did the adults look substantially different? Were scansoriopterygids really tree climbers, and what did they use their long fingers for? And were those unusual ribbon-like structures really used for display? Very little scientific study of these new fossils has been performed, and there is still a great deal to learn.

Right The long tail structures of *Epidexipteryx* may have been boldly patterned or brightly colored. Perhaps the long, projecting teeth were used to grab insects.

Below Perhaps *Epidendrosaurus*, shown here in blue, used its long third fingers to winkle insect larvae out of holes in wood. *Microraptor* leaps past in the background.

A tyrannosaur
named Sue

Tyrannosaurus rex, the most famous dinosaur of all, was also one of the most reported by the end of the 1990s, with about 30 specimens—a large number for a dinosaur that was probably rare in its own day and did not die in groups. Paradoxically, many details of *T. rex*'s anatomy remained poorly known because no major descriptive work had been done since Henry Osborn's in the early 1900s (*see* pp.56–57). Furthermore, arguments over the number of species had been brewing since the 1970s.

Above The gigantic skull of "Sue" the *T. rex*, shown here with fossil collector Peter Larson, has provided a wealth of information on the anatomy, behavior, and sensory abilities of this tyrannosaur.

168

Several people have argued that various fossils represent new *Tyrannosaurus* species. In 1972, Douglas Lawson named *Tyrannosaurus vannus* for a specimen from Texas; in 1988, Robert Bakker and colleagues argued that a tyrannosaurid skull from Montana belonged to a distinct type of dwarf tyrant dinosaur named *Nanotyrannus lancensis*; and in 1995, George Olshevsky named *Stygivenator molnari* for a tyrannosaurid partial skull from a small-bodied specimen, also from Montana. Bakker argued that *T. rex* itself might really be two species (one of which was known simply as *T.* "*x*"). Some paleontologists contested all of these claims, arguing that *Nanotyrannus* and *Stygivenator* were simply juvenile specimens of *T. rex*, and that *T. vannus* and *T.* "*x*" were part of the variation present within *T. rex*. These issues remain the subject of debate even today.

"Sue" is discovered

In 1990, the most complete *T. rex* specimen ever found was discovered by commercial fossil collector Sue Hendrickson in South Dakota. An estimated 73 percent of the skeleton was preserved, and it soon became the focus of intense media interest. Dubbed "Sue," the specimen was excavated by a team from the Black Hills Institute. However, legal issues arose over land ownership. After being seized by the FBI, the specimen was auctioned and eventually purchased by the Walt Disney and McDonald's corporations. Today, it is on display at the Field Museum of Natural History in Chicago.

The specimen's excellent state of preservation enabled a wealth of new paleobiological information to be gleaned from its bones. CT-scanning of its skull revealed the shape of the brain, inner ear, and inside of the nasal cavities. Two large spaces near the front of the brain were identified as olfactory bulbs

(the parts of the brain involved in the perception of odors). Their grapefruit-like size was stupendous. Later study showed that these structures are not the olfactory bulbs after all, and that the real bulbs were far smaller, although still among the largest of any dinosaur.

The inner ear of the specimen shows that tyrant dinosaurs had a highly developed sense of balance and could make precise head and neck movements. Studies of this fossil show that *T. rex* had excellent vision, with good depth perception. Healed wounds and fractures suggest that the dinosaur led a rough life: a healed injury is present on the right upper arm and shoulder girdle, the ribs show signs of healing, and the fibula in the left leg appears to have become infected.

Sexing the specimen

One of the most controversial ideas about "Sue" concerned its sex. Peter Larson and Eberhard Frey suggested that it might be a female because the first chevron appears to be quite small. Chevrons are rod-like bones that project downward from the bottom of the tail vertebrae, and, based on work on modern crocodilians, Larson and Frey thought that females had smaller first chevrons than males to allow easier passage for eggs. In keeping with its vernacular name, "Sue" therefore was referred to as a "she" in many publications.

In 2003, a full description of "Sue" was published by Chris Brochu. At last, *T. rex* was described scientifically in full detail. Brochu argued that the pattern of chevron variation that Larson and Frey reported for crocodilians was not as unambiguous as they stated. He concluded that, due to the disarticulation of the chevrons in "Sue," its gender could not be reliably determined, even if Larson and Frey were correct.

Above "Sue" is now properly known among paleontologists as FMNH PR2081 and is on display in Chicago's Field Museum. Its unusual skull shape is the result of distortion incurred during fossilization.

DISCOVERY PROFILE

Name	*Tyrannosaurus rex* specimen "Sue"
Discovered	South Dakota, United States, by Susan Hendrickson, in 1990
Described	By Chris Brochu, 2003
Importance	The best preserved, most complete *T. rex* specimen discovered, and the first to reveal detailed information on sensory abilities
Classification	Saurischia, Theropoda, Coelurosauria, Tyrannosauridae

Reconstructing dinosaurs

The mounted skeletons of dinosaurs have been the centerpieces of museum collections since the late 1800s, and it is well documented that they have had formative influences on many scientists and artists. Some museums display original fossils in mounted poses, others use replicas made from fiberglass, plaster, or even carved from wood. Because skeletons are frequently incomplete, many mounts incorporate copies of bones or may be composites of two or more specimens. Specially shaped metal and plastic armature, or framework, is used to mount a skeleton. The specimen should be posed and mounted so that the armature is unobtrusive, cleaning is easy, and scientists and others can see the bones well. Many museums have remounted their dinosaurs in recent years to better reflect current thinking.

Right In this museum room, technicians construct the metal armature used to support the skeletons that will be placed on display. The brackets, rods, and pins are often custom-made.

Below left Reconstructed skeletons sometimes provide information on such things as ranges of movement. Here, curator Rolf Johnson uses elastic strips to help estimate forelimb motion possible in a horned dinosaur.

Below right Copies of bones are cast from molds, which are made by painting latex or rubber onto the bones. These rubber molds were used to create a replica *Barosaurus* skeleton.

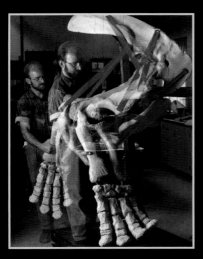

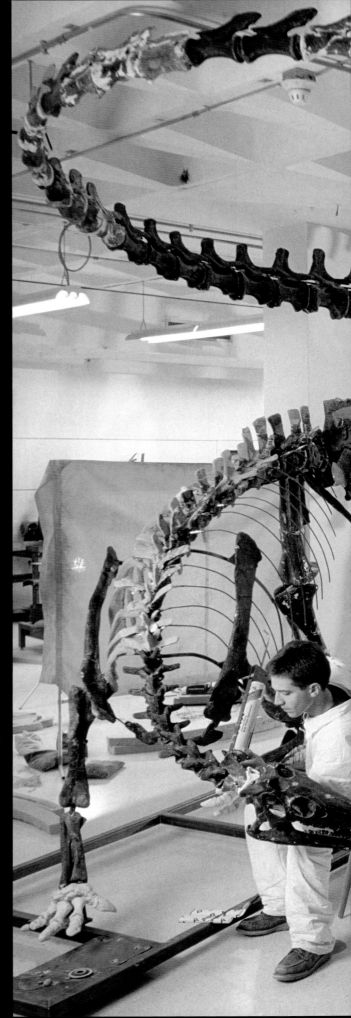

Asian lambeosaurs

Above *Olorotitan* had a longer neck than other lambeosaurine hadrosaurs, and its fan-shaped crest was unique. Its name means "giant swan."

During the last century, the crested hadrosaurs, or lambeosaurines, were regarded as a mainly North American group; only a few fossils were found in eastern Asia. *Tsintaosaurus* from Shandong in China—famous for its bizarre, forward-projecting, spike-like crest—and *Barsboldia*, named in 1981 from Mongolia, were both thought to be lambeosaurines. However, new Asian lambeosaurines, described in 2000 and 2003, showed that Asia was rather more important in the history of these dinosaurs.

Above The Amur region of eastern Russia has proved a rich hunting ground for new hadrosaurs. Here, a team of paleontologists from Belgium excavate a newly discovered species.

The first of these new lambeosaurines was *Charonosaurus jiayinensis*. It was named in 2000 by Pascal Godefroit and colleagues, for a large partial skull that was missing its crest and snout region. The skull was discovered near Jiayin on the southern bank of the Amur River in Heilongjang Province, China. A substantial amount of hadrosaur material was previously collected from this area, and it is likely that several other hadrosaur fossils from this region that have been given different names over the years are the same species as *Charonosaurus*. Because all of the lambeosaurine remains from Jiayin seemed to belong to the same species, Godefroit and colleagues used many specimens to gain the most complete view possible of *Charonosaurus'* anatomy.

Charonosaurus is estimated to have been almost 43 ft (13 m) long and so was a giant; at 49 ft (15 m) only the huge hadrosaurine *Shantungosaurus* was larger. Various features show that *Charonosaurus* was particularly closely related to *Parasaurolophus*, the tube-crested North American lambeosaurine named by William Parks in 1922 (*see* pp.72–73). In a cladistic analysis, Godefroit and colleagues did find these two taxa to group together.

Like *Parasaurolophus*, *Charonosaurus* has particularly robust upper arm bones, but it is unusual in having markedly long, slender lower arm and hand bones. Unfortunately, the crest of *Charonosaurus* is unknown, but the upper surfaces of some of the bones that formed the skull roof show that a large, backward-projecting crest of some sort was present, and it was probably a tubular one similar to that of *Parasaurolophus*.

Olorotitan

A second new Asian lambeosaurine, *Olorotitan arharensis*, was named in 2003. Also named by Godefroit and colleagues, it has a backward-pointing, hatchet-shaped crest unlike that of any other lambeosaurine. It seems to have a particularly long neck compared to its relatives (consisting of 18 vertebrae rather than 15), and a proportionally long shoulder blade. Due to its long neck, its describers gave it a name that means "gigantic swan from the Arhara Country."

Among lambeosaurines, the shape of the snout bones show that *Olorotitan* is a member of the fan-crested clade that includes *Corythosaurus* and *Hypacrosaurus*, and it was perhaps closer to these taxa than to *Lambeosaurus*.

The Tsagayan Group

Olorotitan was excavated between 1999 and 2001 in the Russian Amur Region, and is from a geological unit known as the Tsagayan Group. Two other hadrosaurs—*Kerberosaurus* and *Amurosaurus*—also came from this geological unit, and of these two, *Amurosaurus* was thought to be yet another lambeosaurine. Unlike *Charonosaurus* and *Olorotitan*, however—but like *Tsintaosaurus* and *Jaxartosaurus* from Kazakhstan—*Amurosaurus* was a basal form.

A new view emerged on the evolution and biogeography of these dinosaurs. If *Tsintaosaurus*, *Jaxartosaurus*, and *Amurosaurus* were all basal lambeosaurines, the entire clade must have originated in Asia. Other Asian lambeosaurines were described later: *Nanningosaurus*, named in 2007 from the Nalong Basin in Guangxi, China, appears to be another basal member of the group. *Nipponosaurus*, named in 1936 for a skeleton from Sakhalin Island, was shown in 2004 to be another Asian lambeosaurine.

Land bridge links

The fossil record indicates that, after originating in Asia, lambeosaurines crossed to western North America via the Bering land bridge. By the Campanian stage of the Upper Cretaceous (that is, within just a few million years), North America was home to both tube-crested lambeosaurines, such as *Parasaurolophus*, and fan-crested types, such as *Corythosaurus*, *Hypacrosaurus*, and *Velafrons*. Perhaps members of both of these groups then crossed back into Asia, where they gave rise to *Charonosaurus* and *Olorotitan*. These are both from the Maastrichtian stage of the Upper Cretaceous and are about eight million years younger than most of their North American cousins. However, if anything is clear, it is that the biogeographical history of these dinosaurs is more complex than we realized.

Right A complete crest of *Charonosaurus* has yet to be discovered. However, it is likely that the crest had the same sort of complicated internal anatomy as *Parasaurolophus*, shown here.

Above In life, *Charonosaurus*, shown here, probably looked very similar to its North American relative *Parasaurolophus*. However, it was larger and had much longer hands and arm bones.

Small furry tyrannosaurs

Above *Dilong* is the "tyrannoraptor" that some experts predicted: a small, primitive member of the tyrannosaur lineage with three-fingered hands and a covering of filament-like "proto-feathers."

The great tyrannosaurids of the Late Cretaceous were at first seen as direct descendants of Jurassic megalosaurs and allosaurs. By the end of the 1990s, however, they were identified as members of Coelurosauria, the theropod group that includes the birds and bird-like theropods. Most coelurosaurs are relatively small, and tyrannosaurids are large, so it was inferred that the latter evolved from small ancestors less than about 10 ft (3 m) long. Coelurosaurs are generally long-armed animals that probably used their three-fingered hands to grab prey, so tyrannosaurids presumably behaved in this way early in their evolutionary history.

Below The hilly countryside of northeastern China has become one of the most important dinosaur-bearing sites in the world. The Cretaceous rocks here are famous for their feathered, spectacularly preserved theropods and other fossils.

During the 1990s, American theropod expert Thomas Holtz argued that the most primitive tyrannosaurs were small, relatively long-armed predators that might be imagined as "tyrannoraptors." *Eotyrannus* has obvious indications of a powerful, tyrannosaurid-like bite and also has long arms and long hands (*see* pp.164–165), so it partially fulfills these predictions. But at about 10–13 ft (3–4 m) in length, it is still quite a large animal and not as small as the "tyrannoraptors" that Holtz pictured.

A few other fossils also looked like good early tyrant dinosaurs, including the Jurassic forms *Stokesosaurus* from the United States and *Aviatyrannis* from Portugal, but their remains are fragmentary and little was known of their detailed anatomy.

A "tyrannoraptor"

In 2004, paleontologist Xu Xing and colleagues named *Dilong paradoxus* from the Lower Cretaceous Yixian Formation of Liaoning Province, China. The generic name means

DISCOVERY PROFILE

Name *Dilong paradoxus*

Discovered Lujiatun, Beipiao, western Liaoning Province, China, by a team from the Institute of Vertebrate Paleontology and Paleoanthropology, *c.*2004

Described By Xu Xing and colleagues, 2004

Importance One of the most basal tyrant dinosaurs; the first member of Tyrannosauroidea to have a filamentous covering preserved

Classification Saurischia, Theropoda, Coelurosauria, Tyrannosauroidea

"emperor dragon," and "*paradoxus*" refers to the animal's surprising characteristics. Here at last was a dinosaur that fulfilled Holtz's predictions.

Dilong is small, at less than 6 ft (1.8 m) long, with long arms and long hands, and yet it has many key tyrant dinosaur attributes in the skull and elsewhere in the skeleton. Like *Eotyrannus* and other basal tyrant dinosaurs, *Dilong* appears to be a member of the tyrannosaur group (termed Tyrannosauroidea), but not a member of the short-armed, mostly gigantic tyrannosauroid group Tyrannosauridae. Like other tyrant dinosaurs, the premaxillary bones at the tip of the upper jaw are short, with small, closely packed incisor-like teeth. Its nostril openings are proportionally large, and, on the top of the snout, raised ridges run along either side. These may have helped to reinforce the skull when the animal bit into prey, and other strengthening structures like this are seen elsewhere within the tyrannosauroid clade.

Dilong's neck is long compared to tyrannosaurids, and its three-fingered hands are strikingly long as well. The second finger is robust, whereas the third is extremely slender.

Later tyrant dinosaurs lost the third finger altogether (though they retained the metacarpal in the palm of the hand), but they always kept a robust second finger.

The "crowned dragon"

Additional basal members of Tyrannosauroidea were described soon after *Dilong*. In 2006, Xu and colleagues described a second new taxon, *Guanlong wucaii* (*see* p.159), the generic name of which means "crowned dragon." Unlike *Dilong*, *Guanlong* is Late Jurassic in age. Although it shows a list of characteristic tyrannosauroid features in the skull, it is unique in having a huge, laterally compressed bony crest that projects upward and backward from the top of the snout.

Xu and colleagues inferred that this crest might have been a liability to the animal when it was hunting. Perhaps, they suggested, the flamboyant crest was predominantly for social and sexual display. This is a viable hypothesis, but given that we lack evidence of sexual dimorphism in *Guanlong* (and in other crested dinosaurs, for that matter), it remains highly speculative and requires further investigation.

Above Dark lines can be seen on the sediments around *Dilong*'s bones: they are shown here preserved at the top of the fossil adjacent to tail vertebrae in the center of the lower image. These are filament-like "proto-feathers," which covered the whole body and gave *Dilong* a furry appearance.

Dracorex, the "dragon king"

Fossils of pachycephalosaurs—the bone-headed ornithischians—are rare, and as a result, much uncertainty surrounds their evolution and way of life. However, the naming of an exciting new North American pachycephalosaur in 2006 inspired a vigorous debate about these enigmatic dinosaurs.

Above The *Dracorex* skull is flat, but its head was well decorated with spikes, horns, and bumps. Its narrow snout suggests that it was a selective feeder.

The new fossil was found in the famous Upper Cretaceous sediments of the Hell Creek Formation.

It consists of a near complete, immaculately preserved skull and a few neck vertebrae. The skull bristles with spikes, bumps, and hornlets, and this fantastic appearance led its describers Robert Bakker and colleagues to name the animal *Dracorex hogwartsia*. *Dracorex* means "dragon king," and the species name is for Hogwarts Academy, the fictional school attended by Harry Potter in J. K. Rowling's novels.

A spectacular skull

Dracorex's skull is about 16 in (40 cm) in length. Long, conical spikes project backward from its rear margin, and smaller spikes project in clumps from the snout, from the cheek region, and from the side margins at the back. The teeth are typical for a pachycephalosaur: small with triangular, serrated crowns suited for shredding leaves.

Right Seen here in front view, the skull of *Dracorex* is surprisingly long-snouted. Most of the spikes on its skull are paired, and the large openings at the back of the skull (the supratemporal fenestrae) are also visible.

The absence of an obvious cutting edge along the margins of the tip of the upper jaw led Bakker and colleagues to suggest that *Dracorex* might not have had a beak at the upper jaw tip (as was typical for ornithischians). They thought it might instead have had a thick pad in this region, superficially similar to the one that cows and their relatives have in the same place. This speculation echoes that of Polish paleontologists Teresa Maryańska and Halszka Osmólska in 1974; they proposed that pachycephalosaurs might have had fleshy lips instead of beaks.

What is particularly interesting about the *Dracorex* skull is that it lacks a dome and has large openings on the upper surface of the rear area. These openings, the supratemporal fenestrae, seem to have disappeared as pachycephalosaurs evolved.

Pachycephalosaur family tree

With its unusual combination of skull features, how might *Dracorex* fit into the pachycephalosaur family tree? As a result of detailed work on the ornithischian

DISCOVERY PROFILE

Name *Dracorex hogwartsia*

Discovered South Dakota, by Brian Buckmeier and Pat Saulsbury, *c.*2004

Described By Robert Bakker and colleagues, 2006

Importance A new kind of pachycephalosaur (the first flat-headed form from North America) that raised major questions about the evolution of the group

Classification Ornithischia, Marginocephalia, Pachycephalosauria

family tree, most dinosaur experts think that flat-headed pachycephalosaurs are basal, and that dome-skulled forms, such as *Stegoceras* and *Pachycephalosaurus*, evolved from flat-headed ancestors. With its flat skull roof and large supratemporal fenestrae, *Dracorex* seems to be a basal form. However, its many spikes and skull bumps recall the features of of its better known North American relatives, such as *Pachycephalosaurus*.

Bakker and colleagues argued that existing views on pachycephalosaur evolution could not account for *Dracorex*, and they proposed that the apparently primitive features of *Dracorex* might be secondary reversals, and that it had descended from dome-skulled ancestors. It is true that *Dracorex* looks more like dome-skulled *Pachycephalosaurus* than flat-skulled *Homalocephale*, but it is doubtful whether this is the huge problem that Bakker and colleagues thought.

Aging the specimen

The growth stage of the only known *Dracorex* specimen was also a source of controversy. In 2007, Jack Horner and colleagues argued that *Dracorex* does not represent a distinct taxon, but is a juvenile *Pachycephalosaurus*. It looks different to the adult form, they argued, because these dinosaurs underwent major growth changes as they approached maturity. Support for this hypothesis comes from the fact that we know that substantial changes

in skull shape occurred in the pachycephalosaur's closest relatives, the ceratopsians. If this idea of changing skull anatomy is correct, dome-skulled pachycephalosaurs started their lives with flat skull roofs and with the large supratemporal fenestrae already known to be primitive for the group.

Fights or visual displays

Various features of the *Dracorex* neck vertebrae and skull encouraged Bakker and colleagues to argue that the animal engaged in head-butting fights. Whether pachycephalosaurs butted heads has been the source of argument since the idea first became widespread in the 1970s. In a 1997 article, Ken Carpenter argued that pachycephalosaur neck and skull anatomy was not suited for transmitting forces along the vertebral column. Furthermore, in 2004 Mark Goodwin and Jack Horner showed that ray-like structures present within pachycephalosaur domes were not adaptations for head-butting as was thought; instead, they were temporary structures that disappeared by the time the animal reached adulthood.

Although pachycephalosaurs have often been depicted running toward each other before colliding at speed, some paleontologists now conclude that they engaged in more gentle sparring, perhaps standing next to or in front of opponents. The possibility that domes were not used for butting at all, but instead served as visual display structures, is also considered likely by some authorities.

Above

Pachycephalosaurus differs from *Dracorex* mainly in being less spiky, and in having a tall, thick skull dome. However, these differences might simply be age related.

177

Dinosaur death throes

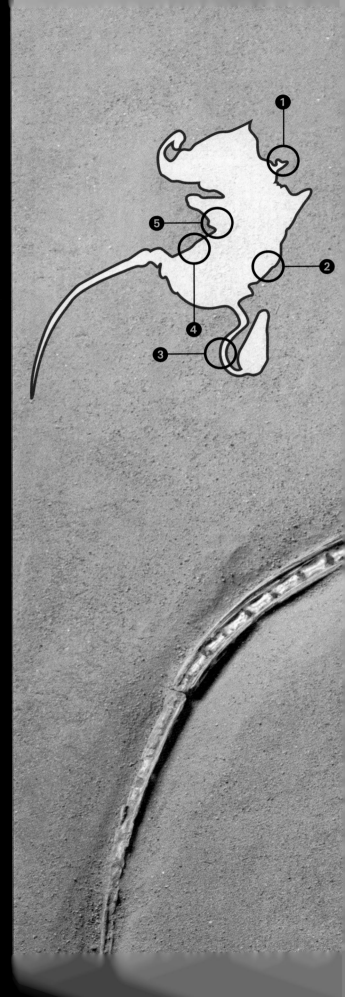

One of the most spectacular fossils ever discovered, the Mongolian "fighting dinosaurs" were locked in combat at the moment of death. Originally encased in sediment, the two skeletons have been fully prepared. Although this fossil shows that *Velociraptor* and *Protoceratops* did sometimes engage in combat, we still do not know how normal this interaction was for these animals.

1 The *Protoceratops* is in a crouching posture, leaning back on its hind limbs and trying to pull itself away from the predator.

2 The narrow, beaked jaws of the *Protoceratops* are clamped tightly shut on the *Velociraptor's* right arm.

3 The *Velociraptor's* neck is strongly curved back so that its head is held well away from the herbivore's powerful jaws.

4 The *Velociraptor's* left foot is pushed up against the *Protoceratops'* neck; its enlarged second toe claw, which was making contact with the large blood vessels in the neck, was probably being used as a stabbing weapon.

5 The *Velociraptor's* left hand lies at the side of the *Protoceratops'* face—it may have been grasping its enemy's head or slashing at its face.

Below It is widely thought that dromaeosaurs such as *Velociraptor* used their giant, sickle-clawed second toe claw as a weapon. The fighting specimen provides direct evidence for this. Here you can see the curved claw pressed against the *Protaceratops'* neck.

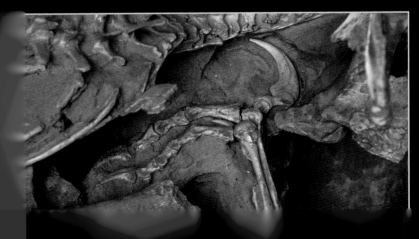

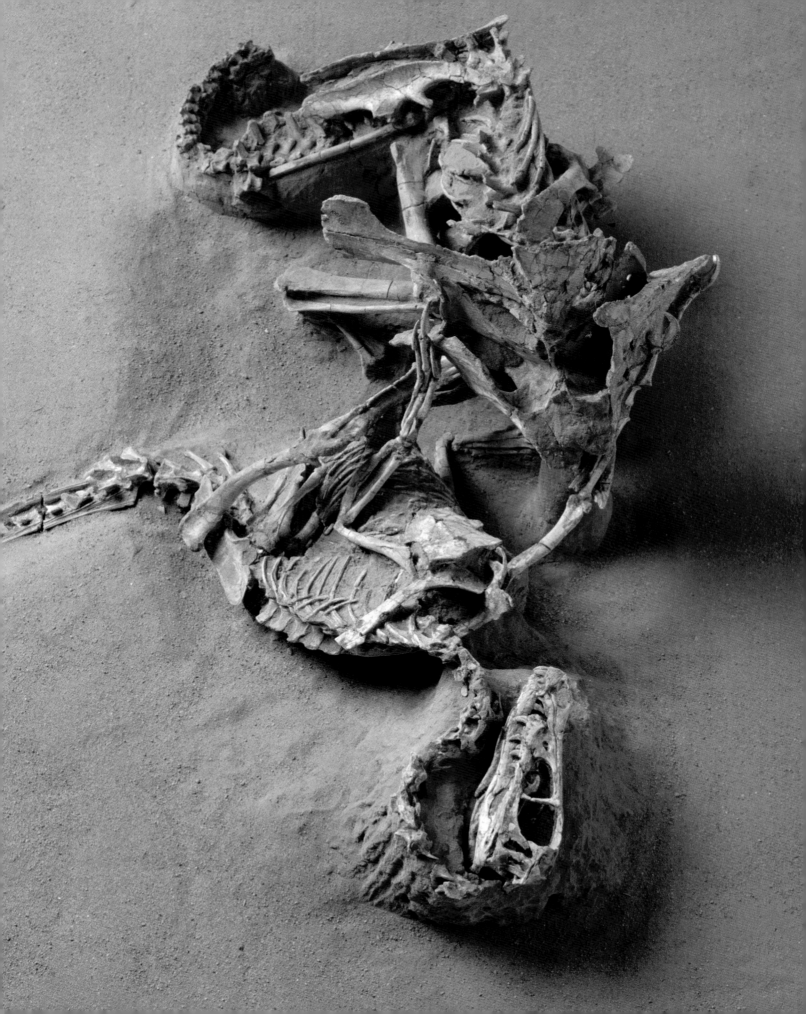

New colossal horned dinosaurs

The description of the new ceratopsids *Einiosaurus* and *Achelousaurus* in 1995 showed that the Upper Cretaceous rocks of North America still had new horned dinosaur species to reveal. And additional new species continued to appear in the early 21st century, helping to fill gaps in the family tree of these amazing animals.

The Chinese paleontologist Xiao-chun Wu of the Canadian Museum of Nature, working with colleagues from the Royal Tyrrell Museum of Paleontology, named the new chasmosaurine ceratopsid *Eotriceratops xerinsularis* in 2007. Like most other chasmosaurines, *Eotriceratops* has long, curved brow horns and a short nose horn. The small bones arranged around the borders of the frill (the epoccipitals) are longer than those of other chasmosaurines, and more crescent-shaped than the semicircular or near-triangular epoccipitals of its relatives. The horn-like bone that projects sideways from the cheek region, called the epijugal, is conical in *Eotriceratops* and thus different from the more irregularly shaped epijugal of other species. The wall-like bony plate located within the nostril opening differs in details from that of other chasmosaurines, too.

Plugging the gap

Unlike *Triceratops*, *Eotriceratops* is not from rocks dated to the very end of the Maastrichtian stage of the Upper Cretaceous. Instead, it is from the upper part of the Horseshoe Canyon Formation, in rocks dated to 67–68 million years ago—and therefore from the middle of the Maastrichtian.

Ceratopsids, including the chasmosaurines *Anchiceratops* and *Arrhinoceratops*, were previously reported from the Horseshoe Canyon Formation, but they were early Maastrichtian in age. *Eotriceratops* was particularly noteworthy in being the first good vertebrate fossil from the upper part of the Horseshoe Canyon Formation and in filling a gap in the chasmosaurine fossil record.

Similarity to *Triceratops*

As an early member of the chasmosaurine group that includes *Triceratops*, *Eotriceratops*, or "dawn three-horned face," has a fitting name. The species name, which means "dry island," refers to the locality in which it was found— Dry Island Buffalo Jump Provincial Park. Like *Triceratops*, *Eotriceratops* is huge, with a skull approximately 10 ft (3 m) long, and it probably looked very similar to *Triceratops* overall.

It is possible that *Eotriceratops* was a direct ancestor of *Triceratops*, but Wu and colleagues also thought that *Eotriceratops* might be the most basal member of a "giant chasmosaurine" clade that includes *Nedoceratops* and *Torosaurus*. *Nedoceratops* has often been regarded as particularly similar to *Triceratops* but Wu and colleagues found that it was more closely related to *Torosaurus*. *Nedoceratops* was long known as *Diceratops*, but in 2007 it was shown that this name was given to a wasp back in 1868.

The skull, horn, and frill shape of horned dinosaurs varied substantially as they matured, so some experts remain skeptical about the validity

DISCOVERY PROFILE

Name *Eotriceratops xerinsularis*

Discovered Dry Island Buffalo Jump Provincial Park, Alberta, Canada, by Glen Guthrie, 2001 (originally discovered, but not collected, by Barnum Brown in 1910)

Described By Xiao-chun Wu and colleagues, 2007

Importance A "missing link" in horned dinosaur evolutionary history

Classification Ornithischia, Marginocephalia, Ceratopsia, Ceratopsidae

of *Eotriceratops*. Its supposedly distinct features, such as the unusually shaped epijugal, might be due to its age. Detailed study is needed to resolve this issue.

More ceratopsid finds

Even if *Eotriceratops* turns out to belong to another, previously named species, it was not the only new ceratopsid species described this century. In 2005, Michael Ryan and Anthony Russell named *Centrosaurus brinkmani* for a new centrosaurine from Alberta. Unlike the other *Centrosaurus* species (*C. apertus*), *C. brinkmani* has an erect nasal horn and small spikes along the top margin of its frill.

Another new species from Alberta, *Albertaceratops nesmoi*, was named in 2007. *Albertaceratops* is also a centrosaurine, but it differs from all previously known kinds in having long brow horns.

It seems to be the most basal known member of its clade, and here at last is evidence demonstrating that centrosaurines started their history with long brow horns. Finally, yet another new centrosaurine from Alberta— the new *Pachyrhinosaurus* species *P. lakustai*—was named in 2008.

Below The gigantic reconstructed skull of *Eotriceratops*, shown here, is very similar overall to that of *Triceratops*. However, the two differ in enough details to show that they are almost certainly distinct.

Glossary

abelisaurid A member of the Cretaceous theropod clade Abelisauridae. Abelisaurids had short arms and short, deep skulls.

acid preparation A technique used to expose fossils from the surrounding rocks. The specimen is soaked in a weak acid solution that slowly dissolves the rock but leaves the fossil intact.

air sacs Sac-like structures distributed throughout the body, and within the bones, of birds and their fossil relatives. Small tubes connect the sacs to the lungs.

allosauroid A member of the tetanuran clade Allosauroidea. The best known allosauroid is *Allosaurus* from the Jurassic of the U.S.A. and Portugal.

ankylosaur A member of the ornithischian clade Ankylosauria, sometimes called armored dinosaurs. All ankylosaurs were herbivorous quadrupeds.

antorbital fenestra An opening on the side of the skull, occupied in life by soft tissues. It is unique to archosaurs but became closed over or reduced in size during the evolution of some clades.

archosaur A member of the reptile clade Archosauria. Archosaurs include crocodilians, birds, and their extinct relatives. They share an antorbital fenestra and other distinctive features.

basal The term used for a taxon that occupies a "low" position on a cladogram relative to other taxa. In the dinosaur cladogram shown on pp. 12–13, Thyreophora is more basal than is Ceratopsia.

binocular vision A type of vision in which the visual fields of the two eyes overlap, providing their owner with enhanced depth perception.

biogeography The study of animal and plant distribution and of the historical factors that led to this distribution.

biped An animal that walks or runs on its hind limbs alone.

braincase The box-like arrangement of bones that surround the brain and partially enclose the nerves and blood vessels associated with the brain.

Cenozoic The period of Earth history that extended from 65 million years ago to the present.

ceratopsian A member of the ornithischian clade Ceratopsia, sometimes called horned dinosaurs. They had bony frills at the back of the skull and a unique bone, called the rostral, in the upper jaw.

ceratopsid The ceratopsian clade that includes the large, horned taxa, such as *Triceratops* and *Pachyrhinosaurus*.

chasmosaurine A member of the ceratopsid clade Chasmosaurinae. Chasmosaurines typically have long snouts and long brow horns.

chevron A Y-shaped bone that projects downward from the underside of a caudal vertebra; sometimes called a hemal arch.

clade A group of organisms that share a single ancestor (and hence is monophyletic). Clades may be huge and consist of hundreds or thousands of species. At the other extreme, they may be subsets of a species.

cladistics The school of thought, which is properly called phylogenetic systematics, in which the branching order of taxa is determined by analyzing the distribution of shared unique features.

coelurosaur A member of the tetanuran clade Coelurosauria. Coelurosaurs possess more slender tails and narrower hands than other tetanurans. The clade includes Tyrannosauroidea, Ornithomimosauria, and Maniraptora.

cold-blooded An informal term, which is properly termed ectothermy, used for organisms that derive their heat from external sources.

Cretaceous The last of the three geological periods that together make up the Mesozoic. It extended from 145 to 65 million years ago.

CT-scanning An imaging technique, which is properly called computed tomography, in which x-rays are used to generate a three-dimensional representation of a structure's interior.

dentary The bone that typically forms the front half or so of the lower jaw (though mammal lower jaws consist entirely of enlarged dentaries).

diplodocoid A member of the sauropod clade Diplodocoidea. Diplodocoids typically have long skulls and long, slender

tail tips. Diplodocoidea includes the clades Rebbachisauridae, Dicraeosauridae, and Diplodocidae.

epoccipital A bone unique to the margins of the ceratopsian frill: it is typically small and semicircular.

family A group of similar genera that typically form a clade. Though once widely used, this and other "ranked" terms are seldom used today.

femur The proper name for the thigh bone: the bone that articulates at its top end with the hip socket and at its lower end with the tibia and fibula.

fibula The more slender of the two long bones in the lower part of the hind limb (the other is the tibia). The fibula is always on the outer side of the limb.

genus A group of similar species that share characters not seen in other related species (plural: genera). Experts often disagree as to which species belong in any given genus.

Gondwana A super-continent, mostly located in the Southern Hemisphere, that resulted from the breakup of Pangaea. Gondwana eventually split to form South America, Africa, Madagascar, India, Australasia, and Antarctica.

hadrosaur A member of the iguanodontian clade Hadrosauridae, though also used for members of the more inclusive clade Hadrosauroidea. Popularly called duckbills or duckbilled dinosaurs.

heterodonty The condition in which an animal possesses teeth that differ in size and shape (the opposite of homodonty). Heterodonty is typical of mammals but is also seen in some reptiles.

iguanodontian A member of the ornithopod clade Iguanodontia. Unlike other ornithopods, iguanodontians lacked teeth in the premaxillae. Among the best known iguanodontians are *Iguanodon* and the hadrosaurs.

infraorder A term once used in dinosaur research for theropod clades that consist of one or two families. An infraorder (such as Oviraptorosauria) is less inclusive than a suborder (such as Theropoda), and a suborder is less inclusive than an order (such as Saurischia). This ranked system of classification is no longer widely used.

inner ear The part of the ear located within the skull and containing the tubes and other structures associated with hearing and balance.

ischium A hip bone that typically projects backward and downward from the pelvic girdle. Although short in mammals, it is typically long in dinosaurs.

Jurassic The second of the three geological periods that make up the Mesozoic. It extended from 199 to 145 million years ago.

lambeosaurine A member of the hadrosaurid clade Lambeosaurinae. Many lambeosaurines had hollow cranial crests.

Laurasia A Northern Hemisphere super-continent that resulted from the breakup of Pangaea. It later split to form Eurasia and North America.

limb girdle The bones that make up the shoulder and hip girdles.

Maastrichtian The last geological stage of the Late Cretaceous. It extended from 70 to 65 million years ago.

maniraptoran A member of the coelurosaur clade Maniraptora. Maniraptorans had particularly long arms and long arm and tail feathers. Maniraptorans include oviraptorosaurs, deinonychosaurs, and birds.

marginocephalian A member of the ornithischian clade Marginocephalia, sometimes called margin-headed dinosaurs. Marginocephalia includes the pachycephalosaurs and ceratopsians.

medullary bone A type of bone laid down by female birds inside the cavities of some of their bones and used exclusively in the manufacture of eggshell. Medullary bone has recently been discovered in various non-avian dinosaurs.

Mesozoic The period of Earth history that extended from 251 until 65 million years ago; it is divided into the Triassic, Jurassic, and Cretaceous periods.

metacarpals (singular: metacarpal) The long bones that form the palm of the hand. Metacarpals articulate with the finger bones at one end and with the wrist bones at the other.

metatarsals (singular: metatarsal) The long bones that form the long part of the foot. Metatarsals articulate with the toe bones at one end and with the ankle bones at the other.

monophyly The condition whereby a group of populations or species descends from a single ancestor. Clades are, by definition, monophyletic.

neural spines The bony spines that (in a quadrupedal vertebrate) project upward from the tops of the neural arches. Neural spines are normally short, but elongate spines evolved in many lineages, including in some dinosaurs.

olfactory bulbs A pair of rounded structures located at the front of the brain and used in the perception of odors. Animals with large olfactory bulbs have a good sense of smell.

opisthopubic condition A configuration of pelvic bones in which the shaft of the pubis extends backward and downward, rather than forward and downward (the propubic condition). This may have evolved so that more room was available for the guts, or so that the center of gravity was farther back in the body. Among dinosaurs, the opisthopubic condition is present in most ornithischians and in birds and some other maniraptorans.

ornithischian A member of the dinosaur clade Ornithischia. They are sometimes called the beaked dinosaurs. Ornithischia includes the thyreophorans, ornithopods, and marginocephalians.

ornithomimosaur A member of the coelurosaur clade Ornithomimosauria, sometimes called "ostrich dinosaurs" or "ostrich mimics."

ornithopod A member of the ornithischian clade Ornithopoda. Among the best-known ornithopods are *Iguanodon* and the hadrosaurids.

orthogenesis A discredited theory, popular during the middle of the 20th century, which supposed that organisms were driven to evolve in a certain direction by an innate "inner force."

ossicles Small plates of bone that occur on or in the skin. The term can also be used for any tiny bone plate: the sclerotic ossicles within the eye are a good example.

pachycephalosaur A member of the ornithischian clade Pachycephalosauria, sometimes called boneheads or boneheaded dinosaurs.

paleontologist A scientist who studies fossils, or uses evidence from fossils to look at patterns in the history of life.

Pangaea The super-continent that existed during the Permian and Triassic when all of the modern continents were conjoined. During the Cretaceous, Pangaea split to form Gondwana and Laurasia.

pelvis A complex assortment of bones located at the posterior end of the body. It includes the three paired pelvic bones and the sacral vertebrae. Also called the pelvic girdle.

"proto-dinosaur" An informal term sometimes used for the closest relatives of Dinosauria within Archosauria. Some, including *Marasuchus*, were bipedal carnivores.

phylogeny The study of relatedness among organisms (properly called phylogenetics). The term is also used for the branching patterns seen within specific clades.

pneumatic Structures that are hollow and filled with air. In biology, the bones and air sacs of birds are pneumatic.

premaxilla A paired bone located at the tip of the upper jaw. The premaxilla typically contains several teeth and forms the lower border to the nostril opening.

prosauropod A name previously used for all those members of Sauropodomorpha that are not part of Sauropoda. Prosauropods were not a clade, and the name is being increasingly abandoned.

pubis The paired bone that, in archosaurs, projects forward and downward from the pelvis. In dinosaurs the pubic bones are typically long (in mammals they are typically short).

quadruped An animal that uses all four limbs in walking or running, rather than the hind limbs alone.

racial senescence A discredited theory, popular in the first half of the 20th century, which proposed that certain lineages somehow ran out of evolutionary potential and then evolved useless, extravagant structures.

radius One of the two long bones that makes up the lower part of the arm (the other is the ulna). The radius articulates at one end with the humerus and at the other with the wrist bones and is always on the same side of the arm as the thumb.

reptile A member of the vertebrate clade Reptilia. The term "reptile" was traditionally restricted to scaly, cold-blooded forms, such as lizards and crocodilians. Used as a clade, Reptilia includes birds and other archosaurs.

saurischian A member of the dinosaur clade Saurischia. Saurischians have longer necks than other dinosaurs and have pneumatic vertebrae. Saurischia includes two main clades, Sauropodomorpha and Theropoda.

sauropod A member of the saurischian clade Sauropoda. Sauropods were generally gigantic, quadrupedal herbivores with long necks and tails.

sauropodomorph A member of the saurischian clade Sauropodomorpha. Compared to other saurischians, sauropodomorphs had particularly long necks.

scapulocoracoid The name used for the largest part of the shoulder girdle, formed from a conjoined scapula (located on the side of the rib cage) and coracoid (located down at the front of the rib cage).

scute A plate of bone, embedded in the skin and covered in life by skin and horn tissue. Scutes were present in early archosaurs and later evolved separately in some dinosaurs.

segnosaur A name previously used for the members of the coelurosaur clade currently known as therizinosauroids.

sexual dimorphism The phenomenon whereby male and female individuals of a species differ from one another in appearance, size, biology, or behavior.

species A population of organisms that are alike and, theoretically, can all breed with one another. The members of a species all share the same diagnostic character or characters not seen in any other similar species.

spinosaurid A member of the theropod clade Spinosauridae. Spinosaurids had long, crocodile-like skulls. Together with megalosaurids, spinosaurids are part of the more inclusive tetanuran clade Spinosauroidea.

squamosal The bone that ordinarily forms the rear upper corner of the skull (in humans it is called the temporal bone). It typically forms part of the edge of the supratemporal fenestra.

stegosaur A member of the ornithischian clade Stegosauria, sometimes called plated dinosaurs. Stegosaurs have rows of tall plates along the neck, back, and tail and long spikes at the tail tip.

sternum A flat bone, typically paired in dinosaurs, present on the ventral surface of the chest. In birds the two sternal plates are fused together.

subadult An animal that is close in size and appearance to a sexually mature individual of its species.

supratemporal fenestrae A pair of openings located on the upper surface of the skull, typically near its rear margin. Jaw muscles are attached to their edges.

taxon (plural: taxa) A group of related individuals that form a species or any other type of clade. The species *Homo sapiens* is a taxon, as are the larger groups Primates, Mammalia, and Vertebrata.

tetanuran A member of the theropod clade Tetanurae. Tetanurans had stiffer tails than other theropods and larger hands. The major tetanuran groups are the spinosauroids, allosauroids, and coelurosaurs.

therizinosauroid A member of the coelurosaur clade Therizinosauroidea—they were once known as segnosaurs. Therizinosauroids had broad hips, long necks, and small heads with beaked jaws.

theropod A member of the saurischian clade Theropoda, sometimes called predatory dinosaurs. All theropods were bipedal dinosaurs with bird-like feet. Theropoda includes the coelophysoids, the ceratosaurs, and the tetanurans.

titanosaur A member of the sauropod clade Titanosauria. Titanosaurs are the largest and most diverse sauropod clade.

Triassic The first of the three geological periods that, together, make up the Mesozoic. It extended from about 250 to about 200 million years ago.

tyrannosauroid A member of the coelurosaur clade Tyrannosauroidea, sometimes called tyrant dinosaurs. Tyrannosauroids had proportionally large skulls.

vertebrae (singular: vertebra) The bones that, together, make up the vertebral column and protect the spinal cord.

vertebrates The groups of animals that have vertebrae around their spinal cord. Vertebrates include fish, amphibians, reptiles, birds, and mammals.

warm-blooded An informal term used for the physiological system, properly called endothermy, practiced by birds, mammals, and various other animals: instead of relying on heat from external sources, they generate their own heat.

Wealden A name used for a Lower Cretaceous rock unit of western Europe, properly called the Wealden Supergroup.

Key figures

Roy Chapman Andrews (1884–1960), American, writer, zoologist. Andrews was based at the American Museum of Natural History in New York. While associated with the Mongolian dinosaur discoveries made during the 1920s, he was predominantly interested in modern mammals.

Robert Bakker (born 1945), American, paleontologist. Bakker studied dinosaur physiology, evolution, and locomotion under John Ostrom during the 1960s. He argued that dinosaurs were successful warm-blooded animals: the ensuing controversy initiated the "dinosaur renaissance." Bakker continues to make controversial claims about dinosaur biology and evolution today.

José Bonaparte (born 1928), Argentinean, paleontologist. Bonaparte has been associated with the discovery and naming of over 25 Argentinean dinosaurs, including *Carnotaurus*, *Mussaurus*, and *Argentinosaurus* (some of which were named with his colleagues or students). His work has revolutionized our understanding of South American dinosaur evolution.

Barnum Brown (1873–1963), American, paleontologist. One of the most famous fossil collectors ever, Brown collected many dinosaurs for the American Museum of Natural History, including *Tyrannosaurus* and *Ankylosaurus*. He described some of these himself and collaborated with William Diller Matthew (1871–1930) and Erich Maren Schlaikjer (1905–1972).

William Buckland (1784–1856), British, geologist, paleontologist, clergyman. Buckland had diverse interests, and published on Pleistocene geology and on the geological evidence for glaciation. His best known contribution to dinosaur research is his 1824 description of *Megalosaurus*. He was an eccentric character who owned many pet animals and also ate an extraordinary variety of exotic creatures.

Alan Charig (1927–1997), British, paleontologist. Charig is best known for his work on Triassic archosaurs as well as on dinosaur origins and evolution. He co-described *Heterodontosaurus* and *Baryonyx*. Due to his position at the

then British Museum (Natural History) and his books, Charig was the popular face of dinosaur research in Britain during the 1970s and 1980s.

Edwin Colbert (1905–2001), American, paleontologist. Colbert is best known for his work on Triassic archosaurs, and for his 1989 description of *Coelophysis*. He was a prolific writer and was one of very few paleontologists writing about dinosaurs during the middle decades of the mid-20th century.

Edward Drinker Cope (1840–1897), American, zoologist and paleontologist. Cope studied living fish, reptiles, and amphibians as well as fossil reptiles and other animals. Best known for his intense rivalry with paleontologist Othniel Marsh, he named *Camarasaurus* and *Coelophysis*, among many others.

Rodolfo Coria (born 1959), Argentinean, paleontologist. An expert on South American theropods, sauropods, and other dinosaurs, Coria has worked with José Bonaparte, Leonardo Salgado, and others in describing such taxa as the gigantic sauropod *Argentinosaurus* and *Giganotosaurus*. His work on the diversity and evolution of titanosaurs (with Jorge Calvo and Leonardo Salgado) has been particularly influential.

George Cuvier (1769–1832), French, naturalist and zoologist. The pioneer of comparative anatomy, Cuvier worked on fossil reptiles and mammals as well as on modern animals. His studies of *Mosasaurus*, ground sloths, and other fossils demonstrated that certain animals had become extinct. Richard Owen and other 19th century scientists aspired to become the "new Cuvier" after his death.

Louis Dollo (1857–1931), Belgian, paleontologist. Dollo supervised the reconstructing and mounting of the *Iguanodon* specimens discovered at Bernissart in Belgium. Unfortunately, he never produced any substantive descriptive work on these fossils, and little information was available on them until the studies of David Norman.

Dong Zhiming (born 1937), Chinese, paleontologist. One of China's most influential paleontologists, Dong has named almost 40 dinosaurs over the course of five decades. He initially worked

under the pioneer of Chinese vertebrate paleontology, Yang Zhong-jian (also known as C. C. Young).

Peter Galton (born 1942), British, paleontologist. A prolific describer of sauropodomorphs, ornithischians, and other dinosaurs, Galton has named numerous dinosaurs and redescribed many that were first named in the late 1800s and early 1900s.

Jack Horner (born 1946), American, paleontologist. Best known for his work on hadrosaur eggs, nests, and babies, Horner also described *Maiasaura* and *Orodromeus*. With colleagues, Horner has worked on the internal anatomy of dinosaur bone and what it means for biology and behavior.

Friedrich von Huene (1875–1969), German, paleontologist. A prolific scientist, Huene wrote numerous articles on the dinosaurs of Europe, South America, Asia, and elsewhere. His best-known works are on Triassic dinosaurs such as *Plateosaurus* and *Thecodontosaurus*, but he also worked on marine reptiles, mammal ancestors, and others.

John W. Hulke (1830–1895), British, surgeon. Although only an amateur geologist and paleontologist, Hulke produced many technical papers on British dinosaurs. He published an important early description of *Hypsilophodon* and also studied *Polacanthus* and *Iguanodon*.

Stephen Hutt (born 1949), British, paleontologist. An expert on the dinosaurs of the British Cretaceous, Hutt has both excavated and described the theropods *Neovenator* and *Eotyrannus*. He has also documented the remains of new *Baryonyx* specimens and key British sauropod specimens.

Thomas Henry Huxley (1825–1895), British, biologist. Huxley—often referred to as "Darwin's bulldog"—was a strong proponent of evolution during the 19th century, and he engaged in debate with Owen and others. Huxley described *Hypsilophodon* and was among the first to argue that birds and dinosaurs are close evolutionary allies. He contradicted Owen's claim that *Archaeopteryx* was much like a modern bird.

Werner Janensch (1878–1969), German, geologist and paleontologist. Janensch is best known for his work on the Jurassic dinosaurs of Tendaguru in Tanzania. He supervised the mounting of these dinosaurs in Berlin's Museum für Naturkunde.

Jim Jensen (1918–1998), American, paleontologist. Sometimes referred to in the media as "Dinosaur Jim," Jensen was an indefatigable collector who excavated numerous dinosaurs. He also pioneered new techniques for the mounting and display of museum specimens. Among Jensen's most notable discoveries are the sauropod *Supersaurus* and the theropod *Torvosaurus*.

Jim Kirkland (born 1954), American, paleontologist. Best known for his work on the Early Cretaceous dinosaurs of the U.S.A., Kirkland has described *Utahraptor* and others. Much of Kirkland's work has focused on revising and describing the ankylosaur fauna of North America and on interpreting the biogeography of North American Cretaceous dinosaurs.

Lawrence Lambe (1849–1934), Canadian, geologist and paleontologist. Lambe is best known for his excavation and description of Cretaceous Canadian dinosaurs. The spectacular dinosaurs he described during the first two decades of the 20th century helped fuel the North American "dinosaur rush."

Wann Langston (born 1921), American, paleontologist. In a career spanning more than five decades, Langston has worked on the theropod *Acrocanthosaurus*, the ceratopsian *Pachyrhinosaurus*, and others. He has also worked on pterosaurs, crocodilians, and other animals. Langston was originally employed to act as Charles M. Sternberg's replacement in Alberta and made numerous discoveries in the field.

Thomas M. Lehman (dates unavailable), American, paleontologist. Lehman is best known for his work on horned dinosaurs from the American southwest; he has also published on the sauropod *Alamosaurus*, and on theropods, turtles and other fossils. Lehman has also examined dinosaur paleobiology.

Joseph Leidy (1823–1891), American, biologist and paleontologist. One of the founding paleontologists of the

United States, Leidy described North America's first dinosaurs (all named for teeth), as well as the far better-represented *Hadrosaurus*. Leidy was also a parasitologist and mineralogist, and he also published on human anatomy.

Richard Lydekker (1849–1915), British, zoologist and paleontologist. Lydekker wrote numerous books and articles on modern animals but is perhaps best known for his many studies of fossil reptiles. However, many of the dinosaurs he named (most of which were from Britain) were based on fragmentary remains, such as isolated vertebrae.

Gideon Mantell (1790–1852), British, medical doctor, geologist, and paleontologist. Mantell described dinosaur teeth in 1825 and named them *Iguanodon* (the name has since been transferred to Louis Dollo's discoveries from Belgium). Throughout the 19th century he described British Cretaceous dinosaurs, and he also published on invertebrates, plants, and other fossils.

Othniel Marsh (1831–1899), American, paleontologist. Pioneering paleontologist who described numerous American dinosaurs from the Jurassic and Cretaceous. He also published on fossil mammals and other animals. Marsh employed field crews who collected many specimens, and he built up a huge collection at the Peabody Museum of Natural History, Yale. Famously, Marsh was great rivals with Edward Cope.

Teresa Maryańska (dates unavailable), Polish, paleontologist. An expert on Mongolian Cretaceous dinosaurs, Maryańska is best known for her work on ankylosaurs, hadrosaurs, and pachycephalosaurs. She published studies on ankylosaurs during the 1970s and, with colleagues, also worked on maniraptoran theropods.

Hermann von Meyer (1801–1869), German, paleontologist. Von Meyer published many classic works during the 1800s on crustaceans, fishes, marine reptiles, and ancient amphibians. In 1859 he coined the name *Archaeopteryx*.

Ralph Molnar (dates unavailable), American, paleontologist. An American who spent some years working in Australia, Molnar described several new Australian dinosaurs. He also published studies on the skull anatomy of *Tyrannosaurus*, on fossil reptile diversity in Australasia, and on fossil crocodilians, pterosaurs, and lizards.

David Norman (born 1952), British, paleontologist. Best known for his work on the ornithopod *Iguanodon*, Norman produced the definitive works on this dinosaur's anatomy and biology. He also worked on ornithischian evolution and feeding biology.

Franz Baron Nopcsa (1877–1933), Hungarian, paleontologist. One of the few paleontologists who published dinosaur research during the 1920s, Nopcsa described many Cretaceous dinosaurs from Romania, and he also wrote about dinosaur evolution and behavior. Nopcsa had a great interest in Albanian culture and made efforts to become King of Albania.

Henry Fairfield Osborn (1857–1935), American, paleontologist. Osborn was associated throughout his career with the American Museum of Natural History in New York and is best known for his work on mammals. He also described dinosaurs, including *Tyrannosaurus* and *Velociraptor*. He held a number of now discredited ideas on the mechanisms of evolution.

Halszka Osmólska (1930–2008), Polish, paleontologist. Osmólska described many dinosaurs from Mongolia. She participated in the joint Polish-Mongolian expeditions of the 1960s and 1970s and described some of the dinosaurs that were discovered on these expeditions, including *Deinocheirus* and *Homalocephale*.

John Ostrom (1928–2005), American, paleontologist. Ostrom described *Deinonychus* during the 1960s; he also published a great deal of important work on dinosaur paleobiology and evolution. His observations on *Deinonychus* prompted him to resurrect the idea that birds evolved from theropods.

Richard Owen (1804–1892), British, comparative anatomist. One of the most influential anatomists of the Victorian Era, Owen worked on both living and fossil animals, but is most famous for naming Dinosauria in 1842. He described many British dinosaurs (many of which had been discovered by Mantell) and was responsible for the 1881 creation of the then British Museum (Natural History).

William Parks (1868–1936), Canadian, geologist and paleontologist. Parks taught geology at the University of Toronto and became an authority on Cretaceous Canadian dinosaurs. During the 1920s he named the Canadian ornithischians *Parasaurolophus*, *Lambeosaurus*, and *Arrhinoceratops*.

Greg Paul (born 1954), American, author, artist, and paleontologist. Famous for his books and illustrations, Paul's distinctive style and attention to anatomical accuracy has inspired a generation of scientific artists, and most accurate renditions of dinosaurs are based on his work. He has also written articles on dinosaur evolution and biology.

Altangerel Perle (born 1945), Mongolian, paleontologist. Best known for his work on the Cretaceous theropods of Mongolia, Perle has worked on therizinosauroids, alvarezsaurids, oviraptorosaurs, and dromaeosaurids. Many of the theropods he discovered were redescribed more recently with American colleagues.

Elmer Riggs (1869–1963), American, paleontologist. Riggs is best known for his work on the giant sauropod *Brachiosaurus*, but he also published studies on other sauropods. Riggs was among the first to argue for terrestrial habits in sauropods, and he even proposed that sauropods such as *Diplodocus* could stand on their hind legs when feeding.

Alfred Romer (1894–1973), American, paleontologist. One of America's most influential paleontologists, Romer is responsible for writing the textbooks that, literally, defined vertebrate paleontology for much of the 20th century. Romer was particularly interested in Triassic archosaurs, but he also studied marine reptiles, mammal ancestors, and ancient amphibians.

Leonardo Salgado (born 1962), Argentinean, paleontologist. Initially working under José Bonaparte, Salgado is one of South America's leading experts on the evolution of sauropods. His work on the diversity and evolution of titanosaurs and diplodocoids has been particularly influential. He has also worked on ornithopods, ankylosaurs, and theropods.

Paul Sereno (born 1957), American, paleontologist. Sereno has published work on primitive ornithischians, sauropodomorphs, ceratopsians, and theropods, and on the evolution and biogeography of dinosaurs as a whole. He led several expeditions to Africa and discovered numerous exciting new dinosaurs, including the sauropods *Nigersaurus* and *Jobaria*.

Charles Hazelius Sternberg (1850–1943), American, fossil collector. Sternberg began his fossil-collecting career under Cope in 1876, but he later worked for the Geological Survey of Canada.

Collecting on contract for various museums around the world, Sternberg excavated many Cretaceous dinosaur specimens. His three sons—George, Charles Mortram, and Levi—continued his line of work, and all became noted collectors and paleontologists.

Charles Mortram Sternberg (1885–1981), American, paleontologist. Son of Charles Hazelius Sternberg, C. M. Sternberg worked for the Geological Survey of Canada and described several Cretaceous ornithischians. He also published work on hadrosaur posture and crest function.

Ernst Stromer von Reichenbach (1870–1952), German, paleontologist. Stromer is best known for the Cretaceous fossils he discovered in Egypt in 1911. They included *Spinosaurus* and various other fossil reptiles—all described by Stromer between 1915 and the mid-1930s.

Philippe Taquet (born 1940), French, paleontologist. Best known for his descriptive work on the dinosaurs of France, Portugal, and Morocco, Taquet has also worked on dinosaur eggs, pterosaurs, and fossil crocodilians. His contribution to north African fieldwork is well known, and his most famous discovery is the Nigerian ornithopod *Ouranosaurus*.

Johann Andreas Wagner (1797–1861), German, zoologist and paleontologist. Wagner specialized on the Jurassic fossils of the Solnhofen Limestone and is best known for his studies of *Compsognathus* and *Archaeopteryx*. He also published on biogeography and on how climate controls the distribution of living things.

Samuel Welles (1907–1997), American, paleontologist. Welles is best known for his studies of *Dilophosaurus* and other Jurassic theropods. From Berkeley in California, he led many fieldwork parties into the desert fossil sites of the American southwest. He also worked on plesiosaurs.

Carl Wiman (1867–1944), Swedish, paleontologist. Wiman is best known for his 1930 description of the Chinese dinosaurs *Tanius* and *Euhelopus*. Wiman also worked on fossil penguins, marine reptiles, and other groups.

Xing Xu (born 1969), Chinese, paleontologist. Xu's work on the dinosaurs of the Lower Cretaceous rocks of China has arguably had more impact than that of any other 21st-century paleontologist. Since 1999 he has (with colleagues) named over 25 new Chinese dinosaurs.

Further resources

BOOKS

Bakker, Robert T. *The Dinosaur Heresies*. New York: Kensington Books, 1995.

Barrett, Paul. *Dinosaurs: A Natural History*. New York: Simon & Schuster, 2002.

Chiappe, Luis M. *Glorified Dinosaurs: The Origin and Early Evolution of Birds*. New Jersey: John Wiley & Sons, 2007.

Colbert, Edwin H. *Dinosaurs: Their Discovery and Their World*. New York: Dutton Books, 1967.

Colbert, Edwin H. *Men and Dinosaurs*. New York: E.P. Dutton, 1968.

Currie, Philip J. and Kevin Padian (Eds). *Encyclopedia of Dinosaurs*. San Diego: Academic Press, 1997.

Czerkas, Sylvia J. and Everett C. Olson (Eds). *Dinosaurs Past and Present, Volume I*. Los Angeles: Natural History Museum of Los Angeles County, 1987.

Czerkas, Sylvia J. and Everett C. Olson (Eds). *Dinosaurs Past and Present, Volume II*. Los Angeles: Natural History Museum of Los Angeles County, 1987.

Desmond, Adrian J. *The Hot-Blooded Dinosaurs: A Revolution in Paleontology*. New York: Dial Press, 1979.

Dodson, Peter. *The Horned Dinosaurs*. Princeton: Princeton University Press, 1998.

Farlow, James O. and Michael K. Brett–Surman. *The Complete Dinosaur*. Bloomington: Indiana University Press, 1999.

Glut, Donald F. *Dinosaurs: The Encyclopedia*. Jefferson: McFarland & Company, 1997

Holtz, Thomas R., Jr. *Dinosaurs: The Most Complete, Up-to-Date Encyclopedia for Dinosaur Lovers of All Ages*. New York: Random House, 2007.

Larson, Peter L. and Kenneth Carpenter (Eds). *Tyrannosaurus rex: the Tyrant King*. Bloomington: Indiana University Press, 2008.

Maier, Gerhard. *African Dinosaurs Unearthed: the Tendaguru Expeditions*. Bloomington: Indiana University Press, 2003.

Martill, Dave and Darren Naish. *Walking with Dinosaurs: The Evidence*. London: BBC Books, 2000.

Norell, Mark, Eugene Gaffney, and Lowell Dingus. *Discovering Dinosaurs: Evolution, Extinction, and the Lessons of Prehistory*. Berkley: University of California Press, 2000.

Norman, David. *The Illustrated Encyclopedia of Dinosaurs*. New York: Barnes & Noble, 1998.

Psihoyos, Louie and John Knoebber. *Hunting Dinosaurs*. New York: Random House, 1995.

Weishampel, David B., Peter Dodson, and Halszka Osmólska (Eds). *The Dinosauria* (second edition). Berkeley: University of California Press, 2007.

Weishampel, David B. and Nadine White. *The Dinosaur Papers 1676–1906*. Washington, D. C.: Smithsonian Books, 2003.

ONLINE INFORMATION

Dinosaurs have a substantial online presence. Unfortunately, very little in-depth information is available and subjects such as dinosaur anatomy, evolutionary history, and the history of dinosaur research have virtually no online presence at the time of writing. The recent (post-2006) explosion of technical blog sites means that blogs are currently the premier source for good, in-depth dinosaur information. Some of the better websites are listed below.

www.dinodata.org/
DinoData is a comprehensive site that provides a huge amount of information on dinosaur taxa, on the Mesozoic world, and on news from the dinosaur world.

www.geol.umd.edu/~tholtz/dinoappendix/appendix.html
This is the online supplement to Thomas Holtz's 2007 book *Dinosaurs: The Most Complete, Up-to-Date Encyclopedia for Dinosaur Lovers of All Ages*. It provides a wealth of information on dinosaur genera and is the most reliable and useful of several similar "dinosaur list" sites.

paleodb.geology.wisc.edu/cgi-bin/bridge.pl
The Paleobiology Database is, as suggested by the name, a huge database of paleontological information. Including contributions from specialists in many areas, it aims to provide a complete listing of sites where fossil species have been found.

www.paleoglot.org/index.cfm
The Polyglot Paleontologist is an invaluable resource for scientists and other researchers. Hundreds of foreign-language paleontological articles have been translated into English, and this site provides free access to them.

www.skeletaldrawing.com/
Skeletal Drawing is the website of Scott Hartman, a paleontologist and artist based at the Wyoming Dinosaur Center in Thermopolis, Wyoming. The site is an excellent resource, providing technically accurate, detailed skeletal reconstructions.

svpow.wordpress.com/
SV-POW!, or Sauropod Vertebra Picture of the Week, is a highly arcane blog site that gathers together technical analyses, photographs, and humor, all of it devoted to the world of sauropod vertebrae. SV-POW! might sound extremely specialized, but it focuses on many issues of great interest to everyone involved in dinosaur research.

scienceblogs.com/tetrapodzoology/
Tetrapod Zoology (the author's blog site) is not devoted to dinosaurs but does cover them in some depth. Lengthy, referenced articles, frequently featuring diagrams and artwork not found elsewhere on the Internet, cover topics such as ceratopsian diversity, Early Cretaceous theropods from England, and the anatomy of sauropods.

SOCIETIES AND ORGANIZATIONS

http://www.dinosaursociety.com/index.html
While the American Dinosaur Society ceased operating some years ago, the British branch of the society remains in action and maintains an active website. An excellent list of updated links to other websites and news stories is maintained.

http://www.vertpaleo.org/
The Society of Vertebrate Paleontology (SVP) is the world's premier organization for professional researchers and scientists involved in the field of vertebrate paleontology. However, interested amateurs are also encouraged to join. The Society produces a technical journal and holds annual meetings.

Index

Page numbers in **bold** denote major treatment of subject. Page numbers in *italic* denote illustrations.

Acknowledgments

I had great fun in writing this book, and I owe a huge debt of gratitude to Deborah Hercun for approaching me with the idea back in December 2007 and for her advice over the following months. Thanks also to Veneta Bullen and Amy Head for their work on images and text, and to Ivo Marloh for his first-rate design work. It has been a pleasure working with Paul Docherty: he provided excellent editorial advice and he and Ben Hoare made many improvements to the text. My thanks also to Mark Goodwin and Charles Crumly for reviewing the text, making numerous helpful suggestions, tightening up the science in places, and correcting dumb mistakes. I'm grateful to Mike P. Taylor and Matt Wedel for their comments on some of the sauropod text, and to all those who provided further advice, assistance, and information: Paul Barrett, Mike Benton, Mike Brett-Surman, Pete Buchholz, Richard Butler, Richard Cifelli, John Conway, Denver Fowler, Peter Galton, Jerry Harris, Thomas Holtz, Dave Hone, Rebecca Hunt-Foster, Steve Hutt, Randy Irmis, Jim Kirkland, Jeff Liston, Leo Salgado, Martin Simpson, Dan Varner, Steve White, and Adam Yates. The etymologies for the dinosaur names were provided by Ben Creisler. Many thanks to those who supplied images (see below), especially Luís Rey, who helped at a relatively late stage of the project. Finally, thanks above all to Toni, Will, and Emma for their love and support.

Every effort was made during the writing of this book to present an unbiased review of the subject and to include those animals, people, and places that I find particularly interesting or "important" in terms of shaping our views. However, it is inevitable that, for reasons of space, many interesting and important discoveries are not discussed and I apologize if my coverage appears unfair or unrepresentative in any way.

Darren Naish Southampton, England, June 2009

PICTURE CREDITS

Marshall Editions would like to thank the following for their kind permission to reproduce their images.   Key t = top   b = bottom   c = center   r = right   l = left

ILLUSTRATIONS

Andrey Atuchin (www.dinoart1.narod.ru): 57t, 82t, 148, 172t, 173t, 174t; **David Bonadonna** (www.davidebonadonna.it): 94b, 101t, 177t; © **Julius T. Csotonyi** (www.csotonyi.com): 64–65; 124, 149b; © Julius T. Csotonyi, courtesy of the Museum of the Red River 2–3, 163t; © Julius T. Csotonyi, courtesy of the Judith River Foundation 102–103; **Steve Kirk**: 10t, 11t, 12–13, 25b, 33b, 36t, 38, 46t, 48t, 58b, 60b, 62t, 67b, 68, 69b, 70b, 75b, 79b, 80–81, 87b, 96t, 104t, 105t, 109b, 112t; **Todd Marshall** (www.marshalls-art.com): 23t, 147, 153, 158, 165; **Luís Rey** (www.luisrey.ndtilda.co.uk): 35t, 41t, 74, 77, 92, 96b, 107, 128, 135, 137, 138l, 140–141, 150b, 155, 167b, 167r, 176.

PHOTOGRAPHS

Pages: 1 Corbis/Louie Psihoyos; 4–5 Corbis/Louie Psihoyos; 6–7 Corbis/Louie Psihoyos; 8 Bridgeman Art Library/Private Collection/© Look and Learn; 9t Corbis/Underwood & Underwood/Bettman Archives; 9b Science Photo Library/Smithsonian Institute; 10–11 Getty Images/Stephen Wilkes/The Image Bank; 15l, 15c, 15r Marshall Editions; 16 Mary Evans Picture Library; 19 FLPA/Norbert Michalke/Imagebroker; 20 Science Photo Library; 21 Scala, Florence/HIP/Science Archive, Oxford 22 public domain; 23b DK Images; 24t public domain; 25t Science Photo Library/Sheila Terry; 27tr DK Images; 28bl Corbis/Martin; Schutt; 28br Corbis/Dung Vo Trung; 28–29 Corbis/Louie Psihoyos; 30b Corbis/Bettman Archives; 30t Oxford University Museum of Natural History; 31 Corbis/DK Limited; 32 Getty Images/De Agostini Library; 33t Academy of Natural Sciences of Philadelphia Library; 34 Corbis/James L. Amos; 35b Corbis/James L. Amos; 36b Corbis/Derek Hall/FLPA; 37t DK Images; 37c National Museum of Natural History/Smithsonian Institution/Paleobiology Department; 37b Bridgeman Art Library; 39 Oxford University Museum of Natural History; 40t DK Images; 40b Natural History Museum, London; 42bl Corbis/Richard T. Nowitz; 42br Ardea/Francois Gohier; 42–43 Corbis/Louie Psihoyos; 44 Corbis/Louie Psihoyos; 45 U.S. Geological Survey; 46b & 47 Corbis/Science Faction/Louie Psihoyos; 48t Alamy/Phil Degginger/Carnegie Museum; 49tl Corbis/Bettmann Archives; 49tr Corbis/Louie Psihoyos; 49b public domain; 50 Corbis/Bettmann Archives; 52 Corbis/Lynton Gardiner/DK Images; 54t Corbis/Samuel Shere/Bettmann Archives; 54b Corbis/Bettmann Archives; 55 Corbis/Lynton Gardiner/DK Images; 56 Corbis/Louie Psihoyos; 56b public domain; 57b Bridgeman Art Library/Private Collection; 58 DK Images/Colin Keates; 59 public domain; 60–61 Corbis/Lynton Gardiner/DK Images; 63 Ardea/Francois Gohier; 66 Corbis/Sandy Felsenthal; 67 Corbis/EPA/Stephanie Pilick; 69t Corbis/Louie Psihoyos; 71 Corbis/Louie Psihoyos/Science Faction; 72 DK Images; 73 Corbis/Encyclopedia; 75t Nopcsa portrait courtesy of the Geological Institute of Hungary ; 76 Corbis/Louie Psihoyos; 78 NHPA/Andrea Ferrar; 79t Corbis/Richard T. Nowitz; 81t Uppsala University; 82b Mathew Wedel; 82–83 Getty Images/Spencer Platt; 84–85 Corbis/Louie Psihoyos/Science Faction; 86 Corbis/Bettman Archives; 87t Ardea/François Gohier; 88 Corbis/Louie Psihoyos; 90 Getty Images/Dorling Kindersley ; 93t Peabody Museum of Natural History, Yale University/Geology Department John Ostrom & Deinonychus skeleton cast; 93b Peabody Museum of Natural History, Yale University; 94t Ardea/François Gohier; 95 Corbis/Kevin Schafer; 97 Hong Kong Science Museum; 98 Corbis/Louie Psihoyos/Science Faction; 99 Corbis/Louie Psihoyos/Science Faction; 100b Corbis/Richard T. Nowitz; 104b Getty Images/De Agostini Picture Library; 105br Dr. Robert Sullivan/Senior Curator, Paleontology & Geology, The State Museum of Pennsylvania, courtesy of the Mongolian Academy of Sciences, Ulaanbaatar; 106 Getty Images/DK Images/Andy Crawford; 109t Corbis/Louie Psihoyos/Science Faction; 110 DK Images; 111 courtesy of Dinosaurland Fossil Museum, Lyme Regis; 112b Corbis/Koen Suyk/EPA; 113 Corbis/Louie Psihoyos; 115 Science Photo Library/Smithsonian; 115–116 Images courtesy of L. M. Witmer & R. C. Ridgely, Ohio University; 116 Getty Images/Dorling Kindersley; 116b DK Images; 117 Getty Images/Dorling Kindersley; 118 Corbis/Louie Psihoyos; 119t public domain, James A. Jensen; 120b National Museum of Canada; 121 Corbis/Paul A. Souders; 122–123 Corbis/Louie Psihoyos; 125 Ardea/Pat Morris; 126 Corbis/Didier Dutheil/Sygma; 129 Queensland Museum; 130t public domain; 130b Brussels Science Institute; 131t Corbis/Louie Psihoyos; 132 Getty Images/Ira Block/National Geographic; 133t & 133r Corbis/Louie Psihoyos; 134 public domain; 136 Ardea/François Gohier; 138–19 Art Archive/Bibliothèque des Arts Décoratifs Paris; 142 Science Photo Library/Mauricio Anton; 143 Vanessa Green; 144 NHPA/Andrea Ferrari; 145 Natural History Museum; 146 & 149t Corbis/Louie Psihoyos; 150t Getty Images; 151 Corbis/Franck Robichon/EPA; 152t Corbis/Philippe Eranian/Sygma; 152b Corbis/Philippe Eranian/Sygma; 154 Ardea/François Gohier; 156 Corbis/Mike Segar/Reuters; 160 Getty Images/Scott Olson; 161 Getty Images/Zhongda Zhang/IVPP; 162t Mike P. Taylor; 162c Mathew Wedel & Richard Cifelli; 162b Mathew Wedel; 163b American Museum of Natural History; 164 Dr. Darren Naish & Robert Loveridge; 166 IVPP v15471A © Professor Fucheng Zhang & Professor Xu Xing, courtesy of Key Laboratory of Evolutionary Systematics of Vertebrates, Institute of Vertebrate Paleontology & Paleoanthropology/Chinese Academy of Sciences, Beijing; 168 Corbis/Layne Kennedy; 169 Corbis/Craig Lovell; 170l Corbis/Louie Psihoyos/Science Faction; 170r DK Images/Lynton Gardiner; 170–171 Corbis/Louie Psihoyos/Science Faction; 172b © Pascal Godefroit/Royal Belgian Institute of Natural Sciences; 173b public domain, Randy Montoya; 174–175 © Professor Fucheng Zhang/Institute of Vertebrate Paleontology & Paleoanthropology, Chinese Academy of Sciences, Beijing; 175 © Mick Ellison; 177b Indianapolis Children's Museum; 178 Corbis/Louie Psihoyos/Science Faction; 178–179 Corbis/Louie Psihoyos/Science Faction; 181 Royal Tyrrell Museum, Alberta.